MOLECULAR CHARACTERIZATION OF PROGESTRONE RECEPTOS IN UTERINE ENDOMETRIUM BY TUMORS

PROF .DR. SAMI AL-MUDHAFFAR

BASMA SALAH AL-OMAR

CHAPTER ONE

Introduction

&

Literature Survey

1.1 THE UTERUS

General description

The uterus (figure 1-1) is a thick-walled muscular organ. This pear-shaped hollow structure is situated in the midline of the female pelvis between the bladder and the rectum[1]. It is about 7.5 cm long, 5 cm wide, and 2.5 cm thick. It is attached to the lateral walls of the pelvis by means of the brood ligaments, anteriorly by the round ligaments, and posteriorly by the uterosacral ligaments, and is covered over most of its surface with peritoneum[2,3].

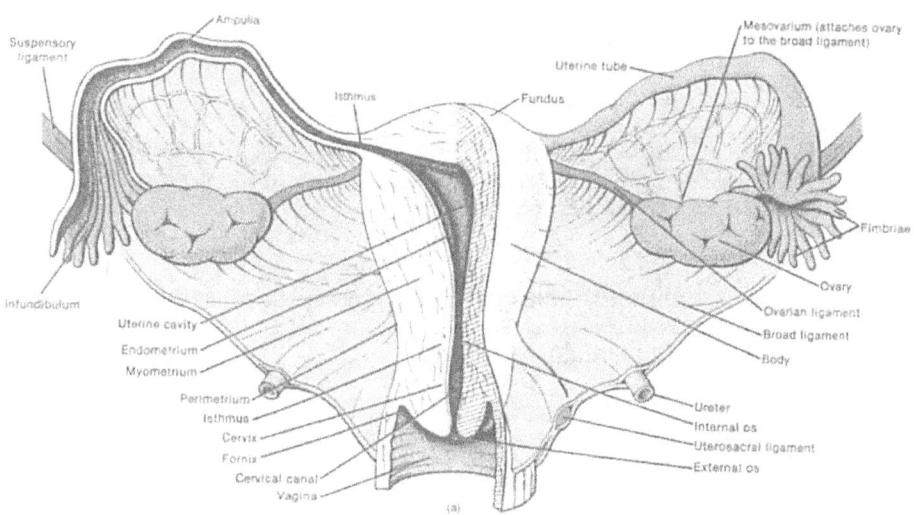

Figure (1-1): The uterus structure

The uterus consists of the following parts[3] :

i. The fundus : The part above the opening of the fallopian tubes.

ii. The cornu : The part into which the fallopian tube open.

iii. The body : The main part of the cavity of the uterus.

iv. The cervix (neck) : The inferior narrow portion opening into the vagina.

v. The isthmus : The narrow lowest part of the body, a constricted region about 1cm long.

The interior of the body of the uterus is called the uterine cavity, and the interior of the narrow cervix is called the cervical canal. The junction of the isthmus with the cervical canal is the internal os. The external os is the place where the cervix opens into the vagina[4].

The wall of the uterus is composed of three layers : the perimetrium, the myometrium, and the endometrium. The perimetrium is the outer serous covering, that is derived from the abdominal peritoneum[1]. The middle muscle layer, the myometrium, forms the major portion of the uterine wall. This layer consists of three layers of smooth muscle fibers and is thickest in the fundus and thinnest in the cervix. The endometrium, the inner layer of the uterus, is made up of a basal and a superficial layer. The superficial layer is shed during menstruation and regenerated by cells of the basal layer[1,4].

1.2 DISEASES OF UTERUS

There are several diseases related to the uterus, such as the Cervicitis, Endometritis and Genital tuberculosis.

1.2.1 CERVICITIS

Cervicitis is an acute or chronic inflammation of the cervix. Acute cervicitis may result from the direct infection of the cervix or may be secondary to a vaginal or uterine infection. The cervix becomes reddened and edematous[5]. Diagnosis depends on taking a swab from the cervical canal[6]. It is treated with appropriate antibiotic therapy. Chronic cervicitis represents a low-grade inflammatory process. The cervix may be ulcerated. Diagnosis depends on vaginal examination, colposcopy, cytologic smear, and occasionally, biopsy to rule out malignant changes. The treatment usually involves cryosurgery[5].

1.2.2 ENDOMETRITIS

Endometritis; this disease can occur as a postpartum infection, with gonococcal, after instrumentation or surgery. Fever and uterine tenderness have been associated with endometritis. The presence of plasma cells is required for diagnosis. Treatment involves either oral or intranvenous antibiotic therapy, depending on the severity of the condition[5].

1.2.3 GENITAL TUBERCULOSIS

Genital tuberculosis is a disease affecting the adult women and occurring mainly between 20 and 40 years. The disease may present with irregular uterine bleeding. Rarely the disease presents in an acute form with fever, night sweat, and loss of weight. Diagnosis depends on finding the mycobacterium tuberculosis in the affected tissues. In most cases examination is made of curettings. Treatment involves the use of antibiotics and chemotherapy should be the first line of attack[7].

1.3 TUMORS OF UTERUS

Tumors of uterus may be either benign or malignant.

1.3.1 BENIGN TUMORS OF THE UTERUS

The benign tumors of the uterus are several such as Leiomyomas, Adenomyosis and Uterine Polyp.

1.3.1.1 LEIOMYOMAS

Uterine Leiomyomas are the most common benign tumors of the uterus, being found in the uteri of about 20% of women over 30 years of age[8]. The peak incidence is between the ages of 35 and 50[9]. The tumor arises from the uterine muscle and grows slowly[10]. They are whiter in colour than the surrounding muscle[11]. Leiomyomas are often multiples, producing an irregular uterine contour. The symptoms associated with uterine lieomyomas are dependent on the size, number, and location of these tumors. Small fibroids are usually asymptomatic. As they enlarge, they may produce symptoms[9]. The most typical symptoms are[12]:

- Excessive menstrual bleeding.
- Urinary problems, constipation, bloating.
- Pain.
- Infertility.

Diagnosis depends on[13]:

- Irregular enlargement of the uterus (may be asymptomatic).
- Dysmenorrhea.
- Acute and recurrent pelvic pain.
- Symptoms due to pressure on neighboring organs (large tumors).

Treatment: choice of treatment depends upon the patient's age, parity, pregnancy status, general health, symptoms, size, location, and state of preservation of the lieomyomas[12].

i. Nonprescription analgesics: (acetaminophen, aspirin, ibuprofen, naprosyn) can treat fibroid- associated pain.

ii. Surgical options consists of myomectomy or hysterectomy.

iii. Medications:

- GnRH agonists are effective in shrinking fibroids, lessening symptoms, and improving anemic conditions. These drugs cannot be used forever, though fibroids begin to grow again 4 to 6 months after treatment with GnRH agonists is ended. For women in their 40s, GnRH agonist treatments may help to avoid surgery by buying time until menopause, when fibroids normally shrink on their own[12].

- Progestogens: It was reported that an intense degeneration change in myomas following large doses of progestogens[14].

- Danazol: (Danol) is a synthetic 2,3-isoxozol derivative of 17α-ethinyl testosterone acts by suppressing gonadotrophin secretion[15]. In 1983 the result of the use of danazol in 8 patients with uterine leiomyoma were published. Although a good hormonal response was obtained following 800mg of danazol daily for 6months in those patients, the reduction observed in the volume of the uterine leiomymas was only 20-25%[16].

- Gestrinone is a synthetic trienic 19-non-steroid, which shares many properties with danazol. It is efficancy in reducing uterine volume of women with leiomyomas is not very impressive at the present time, and seems comparable to the scanty results obtained with danazol[16].

1.3.2 MALIGNANT TUMORS OF THE UTERUS

The malignant tumors of the uterus are several such as Carcinoma of the uterus, Sarcoma of the uterus, Mixed mesodermal tumors and Chorion-carcinoma.

1.3.2.1 CARCINOMA OF THE UTERUS

Two main varieties are recognized

a- Cancer of the Cervix.

b- Cancer of the Uterine body.

a- Cancer of the cervix

Cervical cancer is the most readily detected and, if detected early, is the most easily cured of all other cancers of the female reproductive system. It occurs mainly in the age group 45 to 55 though it may be found in much younger and in older women. It is rare among celibate women. The symptoms begin when the surface of the growth becomes ulcerated and thus symptoms appear late with endocervical growths[5]. The chief symptoms is a watery, often offensive and blood-stained discharge[17].

One of the most important advances in the early diagnosis and treatment of cervical cancer was made possible by the observation that this cancer arises from precursor lesions, which begin with the development of atypical cervical cells. These gradually progress to cancer in situ and finally, to an invasive cancer of the cervix. Cancer in situ is localized to the epithelial layer, whereas invasive cancer of the cervix spreads[5]. Diagnosis depends essentially on biopsy of the cervix which should be combined with curettage and excision of the endocervix[17]. Early treatment of cervical cancer involves removal of the lesion by one of various techniques. Depending on the stage of involvement of the cervix, invasive cancer is treated with radiation therapy, surgery, or both. Both external beam irradiation and intracavitary cesium irradiation (insertion of a closed metal cylinder containing cesium) can be used in the treatment of cervical cancer[5].

b- Cancer of the uterine body (Endometrial Carcinoma)

Carcinoma of the uterine body arises from the endometrium or from an endometrial polyps. Among several risk factors, the most important is age; more than 70% of cases occur in women over 50 years old, compared with only 5% in those younger than 40[18]. Other documented risk factors include obesity, nulliparity, late menopause, hormonally active ovarian tumors, unopposed estrogen supplementation and prolonged use of tamoxifen[19]. As with cervical cancer, it is believed that precancerous abnormalities of the endometrium precede endometrial cancer. These precancerous changes include endometrial hyperplasia or an abnormal pattern of growth in the cells that line the uterus. These cellular changes may be spontaneous, or they may develop secondary to exposure to unopposed exogenous estrogens. Hyperplasia often causes abnormal bleeding and spotting and can be diagnosis with an endometrial biopsy or by dilatation and curettage. The major symptom of endometrial cancer is abnormal, painless bleeding. Any postmenopausal bleeding is abnormal and warrants investigation to rule out endometrial cancer or its precursor stages. Because bleeding is such an early warning sign of the disease.

Late signs of uterine cancer may include cramping, pelvic discomfort, lower abdominal pressure, and enlarged lymph nodes. Dilatation and curettage is the definitive procedure for diagnosis because it provides a more through evaluation[5].

Treatment: The treatment depends upon its pattern of spread.

- Surgical treatment

    Because bleeding is usually an early sign of endometrial carcinoma, most patients present with stage I, grade 1 and 2 disease and can be adequately and completely treated by simple hysterectomy. The abdominal approach is preferred except in patients with very early disease and a small uterus. Radical hysterectomy has been recommended by some, particularly for stage II tumors, but the results have been no better than with simple hysterectomy combined with radiation therapy. Radical hysterectomy can be effective treatment for patients with recurrence following treatment with radiation therapy alone or for those who have previously received therapeutic doses of pelvic radiation therapy for other pelvic cancers. In many stage III patients and most patients with stage IV disease, surgery is not feasible; but when there is reason to expect cure, even pelvic exenteration may be indicated[20].

- Radiation therapy

    Most endometrial cancer patients are candidates for primary surgical therapy with either preoperative or postoperative supplementary radiation therapy. In recent years, many medical centers have tended to withhold any form of radiation treatment until surgical treatment (e.g. hysterectomy, bilateral salpingo-oophorectomy, pelvic and para-aortic node sampling, and procurement of pelvic washing) has been completed and all specimens have been analyzed. Patients are classified on the basis of potential risk of recurrence, and they receive external pelvic irradiation (of whole pelvis therapy) if they are thought to be in the high-risk group. In more advanced stages of endometrial cancer, external irradiation is usually given first, with intracavitary treatment administered later[20].

- Hormonal therapy

Progesterone has been the time-honored agent for the treatment of recurrent endometrial carcinoma not amenable to irradiation or surgery. This type of therapy is relatively free of side effects and can be administered orally or parenterally. Hydroxyprogesterone caproate, other agents such as medroxyprogesterone acetate suspension (Depo-Provera) and megestrol (Megace), appear to have similar effectiveness. The response to progesterone can be quite accurately predicated by levels of estrogen and progesterone receptors in the tumors. Levels of these receptors are inversely proportionate to the grade of the tumor; poorly differentiated (grade 3) lesions have low levels of estrogen and progesterone receptors and usually do not respond to progesterone therapy. While progesterone have an encouraging record in the treatment of recurrent endometrial adenocarcinoma[20]. A statistically significant relationship exists between the presence of cytoplasmic progesterone receptor in recurrent and advanced endometrial adenocarcinomas and the response of these cancers to progestin therapy[21]. New hormonal agents on the horizon include tamoxifen, a potent antiestrogen, which has produced a small number of responses. As with progesterone, the patients who respond generally have well-differentiated tumors and long disease-free intervals[20].

- Antitumor chemotherapy

There are many small studies indicating activity by several agents. Doxorubicin (Adriamycin) is reported to have an approximately 35% overall response rate, with 25% responding completely. Cyclophosphamide (cytoxan), an alkylating agent and fluorouracil (5-FU), an antimetabolite, have both achieved about 25% response rate in isolated small series[20]. Treatment with cisplatin or carboplatin and paclitaxel an active chemotherapeutic agents in endometrial carcinoma[22,23].

1.4 BIOCHEMICAL ASPECTS OF UTERINE TUMORS

1.4.1 TUMOR MARKERS

In order to study the biochemistry of uterine tumor, we have to concentrate on the role of various analytes in the detection and monitoring of these diseases and to explain the phenomenon of uncontrolled growth (malignant and benign)[24]. Tumor markers are used to express the biochemical events of tumors. Tumor markers are substances of different chemical nature that are either produced by a tumor or released by the host response to a tumor[25]. They may be present in the circulation, body cavity fluids or tissues[26]. Disappearance of the marker should indicate eradication of the tumor, whereas an increase in marker concentration should indicate tumor growth[27].

Tumor markers are useful in screening, in determining therapy, in providing prognostic information, in monitoring the response to therapy and in detecting relapse[25]. As diagnostic tools, tumor markers have limitations. The value of a marker depends on its sensitivity, specificity, proportionality and feasibility[28].

Many tumor markers are produced by the tumor cells, their levels would be dependent on the mass of the tumor, and may be influenced by a variety of factors, the number of tumor cells, the proportion of tumor cell synthesizing the marker, the synthetic rate per cell, the location of tumor marker with the cell and the mechanism of release from the cell[29].

Many tumor markers have been investigated in the hope of finding a suitable tool for the detection, diagnosis and management of cancer, but only a few markers have a special position in clinical oncology, due to the difficulties involved in the transfer of technology from the research laboratory to the bedside[30].

Circulated tumor markers can be measured either, chemically or immunologically, i.e. radioimmuno assay, immunoradiometric assay and immunofluorescence assay[31]. Immunological methods could be used for the differentiation of neoplastic cell from normal cell due to their high specificity[32]. Their measurement has been of little value in the detection of asymptomatic cancer. Nevertheless, in some cases they have proved useful in the early detection of local recurrence of malignant disease after treatment, or in detecting the development of metastases[33].

Tumor markers may be characterized as structural or chemical; specific (unique to a single tumor type) or nonspecific (present in a variety of tumor types) (table 1-1)[34].

Table (1-1): Classification of tumor markers

Structural

Submacroscopic (colposcopy)

Microscopic
Ultramicroscopic } membrane
cytoplasm
nucleus

Biochemical

Nonspecific

Tumor: hormones (eutopic or ectopic), fetal proteins, placental proteins, enzymes, nucleic
acid derivatives

Host: acute phase proteins, hydroxyproline.

Specific

Antigens, antibodies and immune complexes

Cell- mediated reactions.

Blocking factors

1.4.2 TUMOR MARKERS OF GYNECOLOGICAL IMPORTANCE

There are several tumor markers which have vary well known in gynecological importance such as

1.4.2.1 FETAL PROTEINS

• *Carcinoembryonic antigen (CEA)*

CEA is a plasma membrane glycoprotein initially discovered in colonic carcinomas, and the systemic levels of CEA were thought to be specific for this cancer[35]. Elevated serum levels of CEA have also been found in patients with carcinoma of the ovary, in particular, mucinous adenocarcinoma, and carcinoma of the cervix, uterus, and breast[36,37]. However, CEA levels are not usually elevated in more than 50% of patients with localized gynecologic malignancies and are less frequently elevated in the early stages of tumor development[38]. Elevation of CEA levels does reflect the tumor volume to some degree, but less precise with large tumor masses[39].

In addition, CEA levels have been useful in the follow-up of ovarian cancer patients to monitor the response of the tumor to chemotherapy[38,39]. However elevated levels of CEA are also found in nongynecologic malignancies, nonmalignant disease' states and healthy individuals[40,41,42]. CEA elevation can be associated with cellular proliferation and may indicate that CEA production is related both to rapid cell division and to specific neoplastic function[43,44].

- *CA-125*

CA-125 is a carbohydrate antigen is present on a variety of normal tissue, including coelomin epithelium and the epithelium of the female reproductive tract and can be detected in 22% of patients with nongynecological cancers and benign gynecological neoplasms[45,46]. Thus it lacks specificity and produces a number of false-positive results[47].

CA-125 is a useful marker in monitoring the progress of nonmucinous ovarian tumors and is invaluable in the management and follow-up of patients. In follow-up, it reflects the response to therapy (surgery, chemotherapy, radiotherapy) in 80% of cases[48,49]. The half-life of CA-125 may also be of prognostic value in that patients with a CA-125 half-life longer than 20 days had a significant increase in tumor progression[50]. However, although CA-125 is useful in the detection of recurrence, it does not detect tumor volumes less than 1 cm[51,52]. Nevertheless, the presence of CA-125 in the patient's serum after therapy, does indicate residual tumor or metastasis[53].

Although CA-125 can be detected in early stages of ovarian cancer, it is not suitable as a screening marker in an asymptomatic population, because normal serum levels of CA-125 do not exclude the presence of other diseases; also, CA-125 is elevated in pelvic inflammatory disease[48,54].

- *CA-15-3*

CA-15-3 similarly, more recent reports of a CA-15-3 indicates that this too may be another useful marker for ovarian, although it has a low specificity since it is elevated in patients with carcinoma of the cervix, endometrium, and vulva[55,56].

Abnormally high serum CA-15-3 levels are found in 32% of endometrial cancer patients. This marker shows a significantly positive correlation with stage of disease, myometrial invasion and tumor differentiation[57]. Nevertheless, CA-15-3 values were correlated with the clinical course of the disease in 87% of ovarian cancer patients[55,56]. This antigen, closely associated with prognostic factors of the disease, may provide clinically useful information on the biological tumor behavior. Moreover, the CA-15-3 assay seems to have a good specificity, since this marker was within the normal range in all patients with endometrial hyperplasia. In contrast, as for other benign gynecological conditions, the CA-15-3 assay shows 6% of false positive[57].

- **Cancer- Associated tumor antigen (CASA)**

  CASA, a new tumor-associated mucin antigen, and most recent studies have shown it to be more sensitive than CA-125 in detecting disease relapse before it was clinically evident[25]. When used in conjunction, serum CA-125 and CASA provided the highest sensitivity[58]. Levels of CASA and CA-125 are raised in peritonitis, chronic liver disease and malignancies of the breast, liver, endometrium and cervix[25].

- **Alpha- Fetoprotein (AFP)**

  AFP is a glycoprotein secreted by the fetal liver, Yolk Sac and gastrointestinal tract. It is raised in the second and third trimesters of pregnancy, in 20-70% of patients with germ cell tumors of testicular, extragonadal and ovarian origin, in 70% of patient with hepatomas[59,60].

  The frequency of increased AFP levels increases with stage of the disease. Serial measurements before, during and after treatment are essential; the half-life of AFP is about 5 days and a deviation from the expected fall in levels with treatment always implies active disease[25]. The presence of AFP in the plasma of significant numbers of patients with gynecologic malignancies may indicate its possible future use as an effective biochemical marker in selected cases[44]. Recently, were reported that plasma levels of AFP were elevated in 30% of patients with gynecologic malignancies. This include patients with ovarian epithelial tumors as well as those with invasive carcinoma of the cervix and endometrium[61].

The use of multiple tumor markers, each of which has moderate specificity for gynecologic cancers, does not seem to provide a significant increase in diagnostic accuracy. Whereas 85% of patients with invasive gynecologic cancer had elevated plasma concentrations of CEA, AFP, or hCG, 31% of the control population also had elevated plasma levels of one or more of these markers[44].

The combination of the three tumor markers carcinoembryonic antigen (CEA), CA-125 and CA-15-3 were allowed the detection of 53% of endometrial cancer patients and 82% of cases at stages II and III. Therefore, as other authors suggest, multiple tumor marker assays could partially compensate for the low sensitivity of a single marker assay. CA-15-3 and CA-125 promise to be clinically useful makers for monitoring response to treatment. Decreasing or increasing levels of CA-15-3 and CA-125 reflected the clinical course of the disease during chemotherapy[57].

Table (1-2): Serum markers in endometrail cancer

Stage	Total Cases	CEA,>4ng/mL		CA-125,>35u/mL		CA-15-13,>30u/mL		Three assay combination	
		n	%	n	%	n	%	n	%
I	30	3(2)	10 (1.5-6)	9 (16)	30 (4-154)	4 (18)	13 (12-48)	11	36
II	9	2(2)	22 (0.5-8)	4 (20)	44 (7-165)	4 (23)	44 (8-85)	6	66
III	8	2(2.2)	25 (1.5-9)	7 (115)	87 (12-500)	7 (66)	87 (12-200)	8	100
Total	47	7	15	20	43	15	32	25	53

Table(1-2) shows the distribution of marker values according to pathological stage of disease. As evident, there was an increasing incidence of abnormal levels of all markers in relation to a higher tumor stage. However, only the percentage of CA-15-3 positively was significantly higher in a more advanced stage (II and III) with respect to stage I[57]. In addition, serum levels of CA-125 measured simultaneously with tissue polypeptide antigen (TPA), CEA, ferritin and some extent CA-19-9, were more useful for early diagnosis, differential diagnosis and early detection of recurrence and remission than CA-125 alone[62,63]. In general, CA-125 levels are commonly elevated in advanced-stage endometrial cancer, although the significance of this finding as an independent prognostic variable has not been established[64,65].

Measuring serial CA-125 levels after primary treatment have not been shown to be reliable in detecting recurrent disease[66].

Other tumor markers, including CA-15-3, lipid-associated sialic acid, CEA, and CA-19-9, are occasionally elevated in endometrial cancer[67]. The value of these tumor markers as prognostic indicators or as a means of monitoring early recurrence following therapy has not been determined[68].

1.4.2.2 ENZYMES

A significant number of enzymes and isoenzymes undergo change in the disarray which accompanies malignant cell transformation[34]. Increased enzymes in serum as well as changes in isoenzyme patterns and newly developed enzyme variants have been found to reflect growth and regression of various malignant neoplasms[69,70].

Several enzymes and their isoenzymes are studied in tumor markers in endometrial cancer patients, such as alkaline phosphatase, and lactate dehydrogenase and others.

Alkaline phosphatase: Heat-stable alkaline phosphatase. Placental alkaline phosphatase (Regan isoenzyme) is normally present in pregnancy and may be regarded as a carcinofetal product, like CEA and AFP. The Regan-like isoenzyme, a heat-stable and L-phenylalanine(phe) sensitive ALP, which had been thought to derive only from cancer or the placenta, was found in uterine cervical reserve cells and endometrial luminal surface lining cells. In contrast, ALP activity in endometrial glandular cells was found to be heat and phe sensitive[71].

Lactate dehydrogenase (LDH; EC 1.1.1.27) in serum consists mainly of five isoenzymes of identical molecular mass but different charge[72]. LDH is a glycolytic enzyme with an increased activity in malignant tissue[34]. Several investigators have reported abnormal electrophoretic patterns of serum LDH isoenzymes and additional bands that correlate with cancers found in humans[73,74].

An extra LDH isoenzyme, designated LDH-1ex, in serum from patients with malignant diseases, that on electrophoresis migrated anodally to LDH-1; this extra isoenzyme was correlated with malignancy[75].

This extra band is the result of an immunoglobulin or β-lipoprotein complex with some of the usual LDH isoenzymes. Reports of finding extra LDH isoenzymes, in sera of patients with lung cancer, colonic cancer, breast cancer, bladder cancer, prostatic cancer, uterine cancer and hepatocellular carcinoma[72].

The activities of, an *acid proteinase*, an *alkaline proteinase*, a *lysine aminopeptidase* were measured in benign and malignant tumors of the human uterus. In carcinomas of the corpus uteri the activity of the acid proteinase was increased compared to normal endometrium. This could probably be the result of cell destruction within the tumor. In leiomyomas of the uterus the activities of the alkaline proteinase and of the lysine aminopeptidase were decreased compared to the normal myometrium[76].

A significant increase in *5α-reductase* activity was found in myoma suggesting that androgens may be involved in myoma pathogenesis[77]. It has been also reported that *estradiol-17 β dehydrogenase* activity, which converts estradiol (E2) to estrone (E1), is higher in the myometrial tissue than in the leiomyoma tissue of the same uterus[78].

Amylase activity was studied in normal endometrium, normal endocervices, endometrial carcinomas and endocervical adenocarcinomas. Amylase was observed in the secretory but not in the proliferative phase of the menstrual cycle. It is possible that the presence of amylase activity may serve a functional role in the degradation of glycogen to glucose in the secretory endometrium. The great majority of uterine cervices showed strong and extensive staining of the endocervical glands for amylase. No glycogen was demonstrated and the role of amylase in endocervical glands remains obscure. Amylase was observed in few cases of endometrial carcinoma and the presence of this enzyme may be considered an eutopic rather than an ectopic expression. Amylase was not detected in any of the endocervical adenocarcinomas examined[79].

Ornithine decarboxylase (ODC) activities were significantly higher in proliferative endometrium during the estrogen-dominated follicular phase of the menstrual cycle than in secretory endometrium after the formation of the progesterone-secreting corpus luteum. ODC activity, were present in both epithelial and stromal cells, may be related to cell proliferation in vivo[80].

Plasma RNase activity was studied in normal and females with gynecological malignancies. The values and abnormal rates raised as clinical stage advanced. Evaluation of plasma RNase value is useful means in biochemical diagnosis and in prognosing gynecological malignancies[81].

1.4.2.3 HORMONES

• *Beta- human chorionic gonadotrophin (β-hCG)*

Human chorionic gonadotrophin (hCG) consists of alpha and beta subunits. The alpha subunit has homology with other glycoprotein hormones such as follicle-stimulating hormone and luteinizing hormone. The beta subunit is larger than the alpha subunit and is immunologically distinct[82]. Beta-hCG (β-hCG) is most commonly elevated in normal pregnancy, in germ cell tumors, in gestational trophoblastic disease, in well differentiated endometrial carcinomas and it was present in the plasma of 22% of the cancer population[27,44].

β-hCG were also elevated in selected cases of epithelial ovarian cancers and immunohistochemical localization of hCG in certain tumors should help to identify those patients in which this marker will be useful clinically[44].

• *Steroid hormones*

Steroid hormones classified into estrogens, androgens, progesterones and corticosteroids, depending on the number of carbon atoms in the molecule and their substituents. They are synthesized through a series of enzymatic reactions starting with a common precursor, cholesterol, which serves as the basic for all steroid hormones. They are water-insoluble compounds that are transported in the blood stream to their site of action by binding to proteins in the plasma.

They are converted by the liver to inactive compounds by a group of hepatic enzymes such as hydrogenases, dehydrogenases and hydroxylases[83].

The steroid hormones are endocrine modulators of metabolism, growth and development[84]. They bind to nuclear receptors, which regulate the transcription of specific genes[85].

Progesterone

Progesterone is a C_{21} steroid hormone secreted as part of the normal menstrual cycle[86]. The corpus lutuem of the ovary secretes large amounts of progesterone after ovulation and in small amounts by the adrenal cortex and the stroma cells of the ovary[15].

Progesterone is an important intermediate in steroid biosynthesis in all tissues that secrete steroid hormones. 17α-Hydroxyprogesterone is apparently secreted from the ovarian follicle. The 20α- and 20β- hydroxy derivatives of progesterone are formed in the corpus lutuem. The progesterone circulates in the blood attached to a specific plasma protein. About 2% of circulating progesterone is free, whereas 80% is bound to albumin and 18% is bound to corticosteroid-binding globulin. It is metabolized in the liver to the major end-product "pregnanediol", which is conjugated to glucuronic acid and excreted in the urine. Progesterone level is approximately 0.9ng/mL (3nmol/L) during the follicular phase of the menstrual cycle. The difference is due to secretion of small amount of progesterone by cells in the ovarian follicles; theca cells provide pregnenolone to the granulosa cells, which convert it to progesterone. During the luteal phase, the corpus luteum produces large quantities of progesterone and ovarian secretion increases about 20-fold. The result is an increase in plasma progesterone to a peak value of approximately 18ng/mL (60nmol/L). The stimulating effect of LH on progesterone secretion by the corpus luteum is due to activation of adenylate cyclase and involves a subsequent step that is dependent on protein synthesis[87].

Mechanism of action of progesterone

The principal target organs of progesterone are the uterus, the breast and the brain[87]. Progesterone is a hormone that induce secretory changes in proliferative endometrium. It has an antiestrogenic effect on the myometrial cells, decreasing their excitability, their sensitivity to oxytocin and their membrane potential. It also decrease the number of estrogen receptors in the endometrium and increases the rate of conversion of 17 β-estradiol to less active estrogens[88]. It also causes retention of water, sodium and nitrogen[15].

The feedback effect of progesterone are complex and are exerted at both the hypothalamic and the pituitary levels. Large doses of progesterone inhibit LH secretion and potentiate the inhibitory effect of estrogens. Progesterone is thermogenic and is probably responsible for the rise in basal body temperature at the time of ovulation. The effect of progesterone, is brought about by an action on DNA to initiate synthesis of new mRNA[87].

Hormone receptors in uterine cancer

Hormones (H) act on cancer cells with specific receptors (R), which, after binding to form a complex (HR), initiate a specific cell response (E) and the general reaction pattern is H+R$\leftrightarrow$HR$\rightarrow$E. Receptor assays in biopsies removed from cancer patients have demonstrated the presence of hormone receptors in almost 60-70% of tumor-sensitive cells and this strongly indicates that these patients have a favorable prognosis and a better response to hormone therapy[89]. Conversely, patients with hormone-negative receptors do not respond to hormone therapy and have a poor prognosis with a greater propensity to metastasis[90,91].

Hormones stimulate DNA and RNA synthesis and oncogene activation. Hormones may stimulate or delay the transformation of precancer cells into cancer cells and also control apoptosis (programmed cell death) of cancer cells. They can also be helpful in developing new anticancer drugs[91].

Estrogen receptor- and progesterone receptor-binding proteins have been demonstrated to be necessary for steroid hormonal function in steroid target and their malignancies[92].

Measurement of estradiol and progesterone receptor (ER and PR) levels may be useful for the prediction of responses to treatment with steroids and antiestrogens. It was suggested that estradiol receptor- and progesterone receptor-positive cancers behaved less aggressively than estradiol receptor- and progesterone receptor-negative cancer[93].

Others have suggested that an absence or low concentration of estradiol receptor and/or progesterone receptor was associated with a poor prognosis in patients with endometrial adenocarcinomas[94]. The correlation between estrogen receptors and endomatrial cancers was studied and found a significantly increased survival time for women with estrogen receptor-positive tumors compared with those who had estrogen receptor-negative tumors[95].

Disease-free survival for both clinical and surgical stage I and II endometrial cancers were significantly better for the progesterone receptor-positive than for the progesterone receptor-negative cancers[96].

Progesterone receptor positive have a greater chance of responding to progestin therapy than progesterone receptor negative[21]. Evidence is now accumulating that patients with receptor-positive tumors have not only a better prognosis, but also, with recurrence, have a greater likelihood of responding to hormonal manipulation[97].

Progesterone receptor in uterine cancer

The presence of receptors in the tissue determines its responsiveness. It seems to us that tissue which respond to steroidal hormones must possess "receptors" which are sterospecific for those hormones-whose combination with the hormone might well be necessary for the sequence of biological and biochemistry events which constitute target tissue responses to the hormone[98].

Progesterone-binding proteins have been demonstrated in several mammalian uterine tissue and in chick oviduct[99,100]. Several investigators characterized these receptors in the uterine of several species including; the mouse, rat, rabbit, guinea pig and human[98,101,102,103]. Others have reported on the physiological and pathological alterations of progesterone receptors in different conditions[92,104].

Wiest and Rao first demonstrated the presence of specific progesterone binding proteins in human uterus. Progesterone interacts with receptors in the human uterus with high affinity and specificity[105]. Different types of tissues such as brain, kidney and breast are responded in somewhat similar manner: i) passage through the target cell membrane, ii) binding to a cytoplasim receptor, iii) activation of the cytoplasmic receptor, iv) transfer of the activated cytoplasmic receptor complex across the nuclear membrane to the nucleus, v) binding of activated cytoplasmic receptor complex to specific sites in nuclear chromatin DNA, vi) synthesis of new mRNA, and vii) synthesis of proteins resulting in modified cellular function[83,106].

The progesterone receptor is an acidic protein with an isoelectric point around pH 5[107]. It has a molecular weight of about 110,000 daltons[104]. Progesterone receptor levels change during the menstrual cycle, in concentration but not in physico-chemical properties. It was showed that progesterone receptor variations were vary large, from a mean value of 6,000sites/cell in the early proliferative phase total receptor increased up to

12,000 sites/cell in the preovulatory period without change of the nuclear to cytoplasmic ratio. After ovulation, there was a decrease in the total receptor content, due to the large decrease of cytoplasmic sites, whereas the level of the nuclear receptor was highest, as might be expected from the transfer of receptor due to the increased concentration of progesterone. At the end of the cycle, the amount of total progesterone receptors was significantly lower than at the beginning[108].

The survival of patients with clinical stage I and II endometrial carcinoma is significantly better for progesterone receptor-positive cancers. Recurrence in patients with stage I disease was significantly more common if tumors were negative for progesterone receptor than if they were positive. Tumors positive for progesterone receptors responded significantly more often than those lacking progesterone receptors[96]. As previously reported on a smaller group of patients, a statistically significant relationship exists between progesterone receptor-positive tumors and response to progestin therapy[21]. This is a consistent observation reported by many investigators[109]. In the cumulative world literature the response rate to progestins for endometrial cancers that are progesterone receptor positive is more than the response rate for progesterone receptor-negative endometrial cancer, thus strongly supporting a role for progesterone receptors in the management of advanced or recurrent endometrial adenocarcinoma[110,111]. Evaluation of receptor content not only allows a better prediction of prognosis, but it also may be helpful in predicting responses to therapy at the time of a future recurrence.

Finally, study of cytoplasmic steroid receptor content allows insight into the cellular regulation of malignancies[92].

AIM OF THE WORK

The aim of the work in this thesis includes the following:

1- Determination of the progesterone level in sera of normal women and patients affected by uterine tumors.

2- Molecular characterization of the binding of ^{125}I-progesterone with receptors in benign and malignant uterine tumors such as those of binding capacity and the effect of various factors (temperature, time, pH, salts, halides, receptor concentration, hormone concentration).

3- Determination of the kinetic and thermodynamic parameters of the binding reaction of progesterone with the receptors of uterine cancer in premenopausal patients.

4- Spectroscopic studies on the progesterone receptors in premenopausal patients with those of benign uterine tumor and malignant uterine tumor of both premenopausal and postmenopausal patients.

CHAPTER TWO

Experimental

2.1 CHEMICALS, INSTRUMENTS AND SAMPLES

2.1.1 CHEMICALS

All laboratory chemicals and reagents were of analar grade and were used without further purification. Tris(Hydroxymethylaminomethan), $MgCl_2$, $MnCl_2$, Dextran, Bovine serum albumin (BSA), were obtained from Fluka.

Dithiothreitol, Gelatin, NaCl, $CuSo_4.5H_2O$, NaF, Na.k.tartarate, Folin cio colteau, were obtained from BDH.

Kit of radioactive progesterone (^{125}I-progesterone) was purchased from INCSTAR corporation (USA). The activity of labeled progesterone was approximately $5\mu ci$.

2.1.2 INSTRUMENTS

The instruments used in this work were, LKB gamma counter type 1270 Rack, LKB spectrophotometer ultraspec type 4050, LKB ultracentrifuge type 2332, pye-Unicam pH meter and Varian DMS 100 UV-Visible Spectrophotometer.

2.1.3 PATIENTS

Two groups of uterine cancer patients and one group of patients with benign uterine tumors were included in this study. Group I contained 8 premenopausal patients with endometrial cancer. Group II consisted of 6 postmenopausal patients with endometrial adenocarcinoma. Group III consisted of 14 premenopausal patients with benign uterine tumors.

All patients were admitted for treatment to Al-Arabae, Al-Saddoon, Saddam Medical City, Yarmook University Hospital, Al-Olwyya University

Hospital under the supervision of specialists Dr.Ryad Mohammad Salh Al-Anni, Dr.Emaad Al-Barack and Dr.Waffa Al-Nassary.

They were histologically proven from the supervision of specialists Dr.Nabbel Al-Sayig, Dr.Luaai Adwar and Dr.Ragy Al-Haddethy, newly diagnosed and not underwent any type of therapy. Patients suffered from any disease that may interfere with our study were excluded.

The host information of all patients and healthy subjects is summarized in (table 2-1).

Table (2-1): The host information of uterine tumor patients and control subjects studied.

Patients	Number	Age (year)	Type of tumors
Group I	8	36.0 ± 5.2	3 patients stage 0 carcinoma is situ. 1 patient stage I tumor limited to endometrium 2 patients stage II invension of more than half of the endometrium width. 2 patients stage IV metastasis (lymph nodes).
Control I	16	38.0± 4.4	
Group II	6	60.0± 1.9	4 patients stage III. 3 patients grade II moderate differentiated. 1 patient grade III poorly differentiated. 2 patients stage IV metastasis (lymph nodes).
Control II	3	62.0± 5.5	
Group III	14	38.0± 5.4	9 patients leiomyoma. 5 patients adenomatous.

2.1.4 PREPARATION OF BLOOD SAMPLES

Blood samples (7mL) were obtained from pre and postmenopausal patients undergoing hysterectomy by venipuncture before operation. Those were left for 20min at room temperature. After coagulation, sera were separated by centrifugation at 3000xg for 10min and kept at −20°C until assaying.

2.1.5 COLLECTION OF SPECIMENS

The tumors tissues were surgically removed from uterine tumor patients by either hysterectomy or myomectomy. The specimens were cut off and immediately rinsed with ice-cold isotonic saline solution. They were collected individually in plastic receptacle and stored at -20°C until homogenization.

2.1.6 PREPARATION OF UTERINE TUMOR TISSUES HOMOGENATE

The frozen tissues were weighed, pulverized finely with a scalpel in petri dish standing on ice bath, and then homogenized at 4°C in buffer solution with a ratio of 1:5 (weight: volume) using a manual homogenizer. The buffer used was Tris-EDTA-dithiothreitol (Tris-HCl 0.01M, pH 7.4, containing 0.0015M EDTA, 1.2mM dithiothreitol and 10% glycerol). The homogenate was filtered through a glass wool in order to eliminate fibers of connective tissues.

The homogenate was then centrifuged at 200xg for 10min at 4°C. The pellet was neglected and the supernatant was centrifuged at 2000xg for 10min at 4°C. The sediment was used to obtain the nuclear fraction and the supernatant was used to isolate the cytosolic fraction. Crude cytosolic fraction was isolated by centrifugation of the supernatant of the cytosolic fraction at

20.000xg for 20min at 4°C. The sediment was discarded and the supernatant was used in the experiments involved cytosolic progesterone receptors .

2.1.7 BUFFERS AND REAGENTS

All buffer solutions were prepared[112] by dissolving the appropriate amount of salts in distilled water and the required pH was adjusted . The stock solution of 0.2M Tris(hydroxymethylaminomethane) was prepared, other reagents were prepared as described previously[113] :

1- TED buffer (pH 7.4): Tris(0.01M, pH 7.4) buffer containing 0.15mM EDTA , 1.2mM dithiothreitol and 10% glycerol .

2- Dextran-coated charcoal (DCC) solution: Tris(0.01M) buffer, pH 7.4 containing 1.25% charcoal, 0.6 % dextran and 0.2% gelatin .

3- Standard unlabeled steroid solution: Stock Solution (250mg/100mL) of the unlabeled steroid hormone was carefully solubelized in a fewest quantity of alcohol then the volume was completed with TED buffer. The desired steroid concentrations were prepared by serial dilution of the initial solution with TED buffer.

2.2 BINDING STUDIES OF ^{125}I-PROGESTERONE WITH THEIR RECEPTORS IN UTERINE TUMOR HOMOGENATE

2.2.1 DETERMINATION OF PROTEIN IN UTERINE TUMOR HOMOGENATE

Protein content of the homogenate of uterine tumors was determined by the Lowry method, using bovine serum albumin (BSA) as the standard protein[114].

Procedure

1- One milliliter of each of standard bovine serum albumin (25,50,100,150, 200µg/mL) was pipetted in a set of duplicate tubes.

2- One milliliter of tumor homogenate was also pipetted in a duplicate tubes.

3- Five milliliter of reagent C was added to all assay tubes.

4- The tubes were shaked and allowed to stand at room temperature for 10min.

5- Half milliliter of reagent D was added to all assay tubes and mixed immediately.

6- The tubes were left at room temperature for 30min.

7- The absorbance of the blue solutions was read at 600nm against the appropriate blank.

8- The standard curve was obtained by plotting the absorbance against the corresponding concentration of standard protein and used to determine the unknown protein concentration of the homogenate of uterine tumors, figure(2-1).

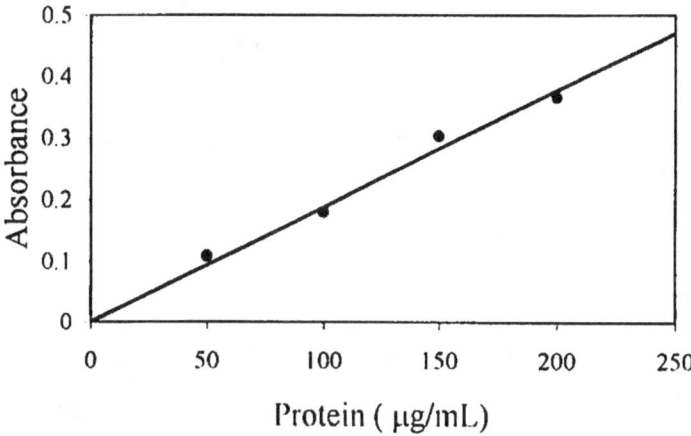

Figure (2-1): The standard curve for protein determination by the Lowry method

Solutions

1- *Reagent A*, Alkaline sodium carbonate solution:

$\qquad$ (2% Na_2CO_3 in 0.1N NaOH).

2- *Reagent B*, Copper sulphate-sodium potassium tartrate solution:

$\qquad$ (0.5% $CuSO_4$.$5H_2O$ in 1% Na, K tartrate).

This solution was prepared freshly by dissolving 0.1gm of Na, K tartrate in 10mL of $CuSO_4$.$5H_2O$.

3- *Reagent C*, Alkaline copper solution. Mix 50mL of reagent A with 1mL of reagent B. Discard after one day.

4- *Reagent D*, Folin Cio calteau reagent: prepared by the dilution of the commercial reagent with an equal volume of distilled water on the day of use.

5- Standard bovine serum albumin (BSA 0.2mg/mL).

2.2.2 DETERMINATION OF ^{125}I-PROGESTERONE CONCENTRATION

The concentration of labeled progesterone was measured according to the method of Morris[115]:

In a set of Rabbit Anti-progesterone coated tubes marked from 1 to 12, 500μl of ^{125}I-progesterone was added with 100μl of standard unlabeled progesterone of different concentrations (0, 0.3, 1, 5, 20 and 60ng/mL). In another set of the same tubes, different volumes (250, 350, 450, 550, 650 and 750μl) of ^{125}I-progesterone were pipetted. After incubation of all tubes for 2h at 37°C, they were decanted and counted using gamma counter.

Calculations

1. (B) is the bound radioactivity (CPM) which represents the counted radioactivity in the precipitated hormone-antibody complex.

2. The free hormone (F) which represents unbound ^{125}I-progesterone was determined from the following formula :

    F (CPM) = Total count (CPM) – Bound radioactivity (CPM)

3. The values of the ratio (B/F) for an ordinary standard curve were plotted against the concentration of standard progesterone.

4. The (B/F) values for the incubation of different amounts of ^{125}I-progesterone were also calculated table(2-2).

5. The data in tables(2-2) and (2-3) were plotted as in figure(2-2).

6. Using the two curves I and II in figure(2-2) we can get the amount of radioactivity corresponding to the concentration of unlabeled hormone table(2-4). This was done by drawing a line which intersects with both curves at the same increase as shown in figure(2-2).

7. The amount of the standard progesterone was plotted against the amount of the corresponding radioactivity table(2-4) and figure(2-3).

8. The concentration of the ^{125}I-progesterone was determined from the intercept of the straight line as in figure(2-3).

Table (2-2) : B/F values corresponding to different concentration

of standard progesterone used in standard curve

Concentration of standard (ng/mL)	B/F
0	0.555
0.3	0.493
1	0.357
5	0.266
20	0.170
60	0.111

Table (2-3) : B/F values corresponding to different amounts

of ^{125}I-progesterone used in the incubation

Amount of ^{125}I-progesterone (CPM)	Bound radioactivity (CPM)	B/F
63546	22300	0.540
82445	27150	0.491
101213	31411	0.452
115373	35364	0.443
154029	40226	0.351
214225	47123	0.282

Table(2-4) : The mass of the standard progesterone in ng/mL

corresponding to a given amount of the tracer

Bound radioactivity (CPM×10³)	Amount (ng/mL)
40	1
42	1.5
43	2.5
45	4
47.5	5

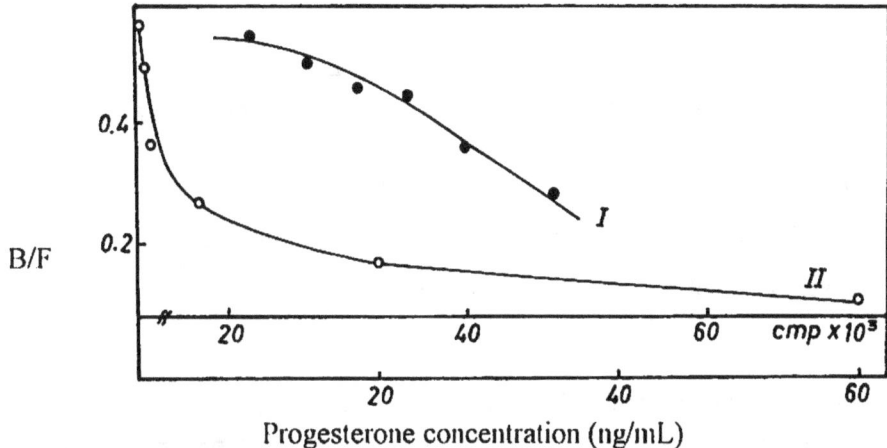

Fig(2-2): Ratio of bound to free radioactivity for an ordinary standard curve, where different amounts of standard progesterone were incubated with constant amount of ^{125}I-progesterone and antibody (II) or, antibody incubated with different amounts of ^{125}I-progesterone in the absence of unlabeled progesterone (I).

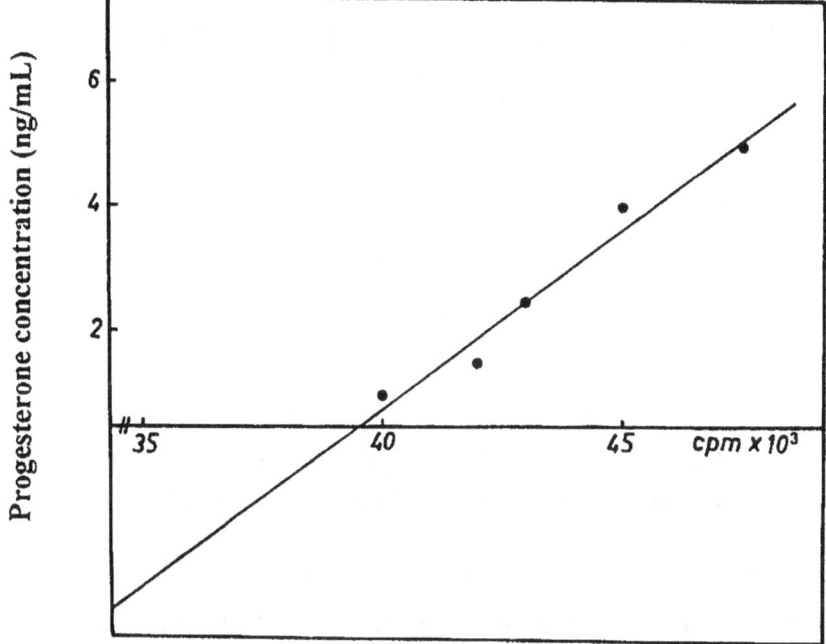

Fig(2-3): A plot of the mass of standard progesterone against CPM of ^{125}I-progesterone having the same B/F values from figure (2-2), which resulted a straight line.

2.2.3 DETERMINATION OF PROGESTERONE LEVELS IN SERA OF UTERINE TUMOR PATIENTS

Serum progesterone levels were measured on samples collected from individuals of pre and postmenopausal patients by radio immuno assay (RIA). The assay protocol was described in table(2-5).

Table(2-5): RIA assay protocol of serum progesterone (ng/mL)

	Progesterone standards (ng/mL)						Unknown	
	0	0.3	1	5	20	60	1	2
Tube No.	1.2	3.4	5.6	7.8	9.10	11.12	13.14	15.16
Standard serum	100	100	100	100	100	100	-	-
Unknown serum	-	-	-	-	-	-	100	100
^{125}I-progesterone	500	500	500	500	500	500	500	500

The test tube rack was mixed by shaking gently by hand and all tubes then incubated in a water bath at 37°C for 60-70min. Then the tubes were aspirated and counted in gamma counter for one minute.

2.2.4 PRELIMINARY TEST OF ^{125}I-PROGESTERONE BINDING TO ITS RECEPTORS IN BENIGN AND MALIGNANT UTERINE TUMOR

Cytosolic receptors was detected using two sets of experiments. The first one was carried out to determine the total binding, while the second was used for the estimation of non-specific binding. In order to detect cytosolic receptors, 200µl (250µg protein) of crude cytosol was incubated with 40µl of labeled progesterone in duplicate tubes. The volume of the mixture was completed to 1mL with TED buffer (pH 7.4), then the tubes were incubated for 16h at 4°C. Non-specific binding was accounted by preparing the same incubation mixture with the addition of 200 fold excess of unlabeled progesterone as competitor. At the end of the incubation (16h), bound and unbound hormone were separated by charcoal absorption with a suspension of dextran-coated charcoal[113]. For this purpose 250μl of dextran coated charcoal-Tris buffer solution was added. The tubes were shaken for 10min and centrifuged for 10min at 2000xg at 4°C. 1mL was taken from each supernatant and counted in gamma counter. It represents the bound progesterone (^{125}I-progesterone).

Solutions

TED buffer prepared as described previously by dissolving the following compounds: 1.25gm charcoal, 0.6gm Dextran and 0.2gm gelatin in 100mL of Tris (0.01M) buffer, pH 7.4.

Calculations

1- The counted radioactivity in each tube (expressed in CPM) represents the total binding (TB).

2- The counted radioactivity (expressed in CPM) in the tubes contained labeled hormone and excess of unlabeled hormone represents the non-specific binding (NSB).

3- The specific binding (CPM) was calculated by subtracting the radioactivity (CPM) obtained in the presence of unlabeled hormone from that produced in the absence of unlabeled hormone.

$$SB (CPM) = TB (CPM) - NSB (CPM)$$

4- The percent of specific binding (SB%) can be calculated from the following formula:

$$SB\% = \left(\frac{SB}{TC}\right) \times 100$$

Where

SB : specific binding (CPM) and

TC : Total concentration of hormone (CPM).

2.2.5 RADIO RECEPTOR ASSAY STUDIES OF ^{125}I-PROGESTERONE BINDING TO ITS RECEPTORS IN BENIGN AND MALIGNANT UTERINE TUMOR

All the following experiments were carried out with two sets of duplicate tubes. The first one was used to estimate the total binding and the second to estimate the non-specific binding .

2.2.5.1 THE EFFECT OF DIFFERENT CONCENTRATIONS OF ^{125}I-PROGESTERONE ON THE BINDING WITH ITS RECEPTORS IN UTERINE TUMOR HOMOGENATE.

Increasing concentrations (0.796×10^{-8} - 7.165×10^{-8} M) of ^{125}I-progesterone was each added to 200µl (250µg protein) of crude cytosol in the first set of tubes with a final volume of 1mL (completed with TED buffer). The second set of tubes consists of the same reactants plus 200 fold excess of unlabeled progesterone. After incubation for 16h at 4°C , 250µl of DCC was added in order to estimate the bound hormone. Then the tubes were shaken for 10min and centrifuged for 10min at 4°C at 2000xg. 1mL was taken from each supernatant and counted in gamma counter . It represent the bound hormone.

Solutions

All solutions prepared as described previously in sections (2.1.7) and (2.2.4).

Calculations

1- The percent of specific binding (SB%) was determined according to the following formula

$$SB\% = \left(\frac{SB}{TC}\right) \times 100$$

Where

SB: Specific binding (CPM) and

TC: Total concentration of hormone (CPM).

2- The percent of specific binding (SB%) was plotted against the concentration of ^{125}I- progesterone.

2.2.5.2 THE EFFECT OF PROGESTERONE RECEPTOR CONCENTRATION ON THE BINDING IN UTERINE TUMOR HOMOGENATE

Forty micro-liter (20ng) of ^{125}I-progesterone was added to 200µl of increasing amounts (50, 100, 150, 200, 250µg) of crude cytosol in a final volume of 1mL (completed with TED buffer) with or without the addition of 200 fold excess of unlabeled progesterone. At the end of incubation (16h) at 4°C, the bound hormone was estimated by adding 250µl of DCC, then the tubes were shaken for 10min and centrifuged for 10min at 4°C at 2000xg. 1mL was taken from each supernatant and counted in gamma counter. It represent the bound hormone.

Solutions

All solutions prepared as described previously in sections (2.1.7) and (2.2.4).

Calculations

1- The percent of specific binding (SB%) was determined according to the following formula:

$$SB\% = \left(\frac{\text{Specific binding (CPM)}}{\text{Total concentration of hormone (CPM)}} \right) \times 100$$

2- The percent of specific binding (SB%) was plotted against the amount of protein receptors included in each mixture.

2.2.5.3 THE EFFECT OF TEMPERATURE ON THE BINDING OF ^{125}I-PROGESTERONE TO ITS RECEPTORS IN UTERINE TUMOR HOMOGENATE

Forty micro-liter (20ng) of labeled progesterone was added to 200µl (200µg protein) of crude cytosol in a final volume of 1mL (completed with TED buffer) with or without the addition of 200 fold excess of unlabeled progesterone. After incubation for 16h at 4°C, the bound hormone was estimated by adding 250µl of DCC, then the tubes were shaken for 10min at 4°C at 2000xg. 1mL was taken from each supernatant and counted in gamma counter, the experiment was repeated at different temperatures (10, 25, 37, 40°C).

Solutions

All solutions prepared as described previously in sections (2.1.7) and (2.2.4).

Calculations

1- The percent of specific binding (SB%) was determined according to section(2.2.4) at each temperature.

2- The percent of specific binding (SB%) was plotted against the different temperatures of incubation.

2.2.5.4 THE TIME-COURSE OF RECEPTOR BINDING IN UTERINE TUMOR HOMOGENATE

Forty micro-liter (20ng) of ^{125}I- progesterone was added to 200µl (200µg protein) of crude cytosol, with or without the addition of 200 fold excess of unlabeled progesterone. The volume of mixture was completed with TED buffer to 1mL. The tubes were incubated at 25°C. At different time intervals (2, 4, 8, 16, 18h). At the end of incubation, the bound hormone to cytosolic receptors was estimated as described in section (2.2.4).

Solutions

All solutions prepared as described previously in sections (2.1.7) and (2.2.4).

Calculations

1- The percent of specific binding (SB%) was determined according to section (2.2.4) at each time.

2- The percent of (SB%) was plotted against the different times of incubation.

2.2.5.5 THE EFFECT OF pH ON THE RECEPTOR BINDING IN UTERINE TUMOR HOMOGENATE

Crude cytosolic fractions (200µg protein in 200µl) were added to 40µl (20ng) of labeled progesterone with or without the addition of 200 fold excess of unlabeled progesterone. The volumes of the mixtures were made up to 1mL with TED buffer of different pH (7.2, 7.4, 7.6, 7.8, 8.0). The tubes were incubated at 25°C for 16h. After incubation, the bound hormone was estimated as mentioned in section (2.2.4).

Solutions

All solutions prepared as described previously in sections (2.1.7) and (2.2.4).

Calculations

1- The (SB%) was estimated as mentioned in section (2.2.4) at each pH.

2- The percent of specific binding was plotted against their corresponding pH.

2.2.5.6 STABILITY OF ^{125}I-PROGESTERONE-RECEPTOR COMPLEX

This experiment was carried out at the optimum conditions of labeled progesterone concentration (6.369×10^{-8} M), protein receptor concentration (200µg), temperature (25°C), time of incubation (16h) and pH (7.6), in order to investigate the effect of temperature on receptor properties. The experiment was performed by adding 40µl of ^{125}I-progesterone to crude cytosolic fraction (200µg protein in 200µl) with or without the addition of 200 fold excess of unlabeled progesterone in a final volume of 1mL (completed with TED buffer pH 7.6). The tubes were incubated at 25°C for 16h. After incubation, the bound hormone, the hormone-receptor complex was evaluated. After the evaluation of the bound hormone, the hormone-receptor complex was reincubated at different temperatures (0, 25, 45°C). Between 0 and 8h the remaining bound hormone in each tube was measured by gamma counter.

Solutions

All solutions were prepared as described previously in sections (2.1.7) and (2.2.4).

Calculations

1- The (SB%) was estimated as mentioned in section (2.2.4).

2- The percent of specific binding (SB%) was plotted against the time of incubation.

2.2.5.7 COMPETITIVE EFFECT OF DIFFERENT CONCENTRATIONS OF UNLABELED ESTRIOL AND PROGESTERONE ON THE BINDING OF ^{125}I-PROGESTERONE TO ITS RECEPTORS IN UTERINE TUMOR HOMOGENATE

The experiment was carried out at the optimum conditions of labeled progesterone concentration (6.369×10^{-8} M), protein receptor concentration (200µg), temperature (25°C), pH(7.6), and the time of incubation (16h). The experiment was performed by adding 40µl (20ng) of labeled progesterone to 200µl (200µg protein) of crude cytosol fraction with or without the addition of increasing concentration (1-1000µmole) of unlabeled progesterone in a final volume of 1mL (completed with TED buffer). After incubation for 16h at 25°C, the bound hormone was measured as described in section (2.2.4). The experiment was repeated with increasing concentrations of unlabeled estriol.

Solutions

All solutions were prepared as described previously in sections(2.1.7) and (2.2.4).

Calculations

1- The percent of specific binding was estimated as mentioned in section (2.2.4).

2- The (SB%) was plotted against the concentration of competitor (progesterone or estriol).

2.2.5.8 EFFECT OF DIFFERENT HALIDES ON THE BINDING OF ^{125}I-PROGESTERONE TO ITS RECEPTORS

Forty micro-liter labeled progesterone (20ng) was added to 200µl (200µg protein) of crude cytosol fraction with or without the addition of 200 fold excess of unlabeled progesterone in a final volume of 1mL (completed with TED buffer containing 0.01M of each of the following halides: NaF, NaI and NaCl, pH 7.6). The tubes were incubated for 16h at 25°C, then the bound hormone was estimated as mentioned in section (2.2.4). Used a sample without the addition of any halide as a control.

Solutions

Halide solutions prepared in concentration of 0.01M in TED buffer pH 7.6, 0.3975gm of NaI in 250mL TED buffer, 0.1463gm of NaCl in 250mL of TED buffer, and 0.105gm of NaF in 250mL of TED buffer.

Calculations

1- The (SB%) was estimated as mentioned in section (2.2.4) at each halide.

2- The percent of specific binding was plotted against halide concentration.

2.2.5.9 EFFECT OF DIVALENT CATIONS ON THE BINDING OF ^{125}I-PROGESTERONE WITH ITS RECEPTORS IN UTERINE TUMOR HOMOGENATE

Crude cytosolic fractions (200µg protein in 200µl) were added to 40µl (20ng) of labeled progesterone with or without the addition of 200 fold excess of unlabeled progesterone in a final volume of 1mL (completed with TED buffer containing 25mM of each of the following salts: $MgCl_2$, $MnCl_2$, $CuSO_4.5H_2O$, pH 7.6). The tubes were incubated for 16h at 25°C, the bound hormone was measured as described in section (2.2.4). Used a sample without the addition of any salts as a control.

Solutions

The stock solution (25mM) of divalent cations were prepared as the following: 0.1489gm $MgCl_2$ in 250mL of TED buffer pH 7.6, 0.1967gm $MnCl_2$ in 250mL of TED buffer pH 7.6 and 0.3898gm $CuSO_4.5H_2O$ in 250mL of TED buffer pH 7.6.

Calculations

1- The (SB%) was estimated as mentioned in section (2.2.4) at each salt concentration.

2- The percent of specific binding (SB%) was plotted against salt concentration.

2.2.5.10 DETERMINATION OF THE CONCENTRATION OF PROGESTERONE RECEPTORS AND THE AFFINITY CONSTANT OF ^{125}I-PROGESTERONE ASSOCIATION WITH ITS RECEPTORS IN UTERINE TUMORS

Cytosolic receptors were measured by the addition of increasing concentration (4.777×10^{-8} - 9.554×10^{-8} M) of labeled progesterone (^{125}I-progesterone) to 200µl (200µg protein) of crude cytosol with or without the addition of 200 fold excess of unlabeled progesterone in a final volume of 1mL (completed with TED buffer). The tubes were incubated at 25°C for 16h, the bound hormone was estimated as mentioned in section (2.2.4).

Solutions

All solutions prepared as described previously in sections (2.1.7) and (2.2.4).

Calculations

1- The values of bound (B) and free hormone (F) were determined as in section (2.2.4).

2- The values of ^{125}I-progesterone which is bound specifically in molar were calculated using the following formula:

$$B_{specific} = \left(\frac{\text{Total binding - Nonspecific binding}}{\text{Total count}} \right) \times \frac{\text{Concentration of hormone}}{\text{in each assay tube}}$$

3- The concentration of receptors and the affinity constant were determined according to Scatchard equation[116].

$$\frac{B}{F} = \frac{1}{Kd} \times (B_{max} - B)$$

$$Ka = \frac{1}{Kd}$$

Where

B: the concentration of bound hormone specifically

F: the concentration of free hormone

Ka: the affinity constant

B_{max}: the maximal binding capacity

Kd: the dissociation constant

4- The values of the ratio B/F was plotted against the values of the B, the receptor concentration and the affinity constant were calculated from the x-axis and the slope of the straight line respectively.

2.2.6 THE KINETIC AND THE THERMODYNAMIC STUDIES

2.2.6.1 THE TIME-COURSE OF ^{125}I-PROGESTERONE BINDING TO ITS RECEPTORS IN PRE MENOPAUSAL PATIENTS WITH UTERINE CANCER

1. At zero time, 40µl of ^{125}I-progesterone (original conc. 1.59µM) was incubated with 200µl (200µg protein) of uterine homogenate. The final volume (1mL) was made up by adding the assay buffer (0.01M TED buffer pH 7.6). The reaction mixture was incubated at 25°C for several time intervals (2, 4, 8, 16, 18h).

2. After incubation 250µl of DCC were added in order to estimate the bound hormone.

3. The tubes were shaken for 10min and centrifuged for 10min at 2000xg at 4°C.

4. One milliliter was taken from each supernatant and counted in gamma counter, It represent the bound hormone.

5. Parallel experiments were performed to determine the amount of non-specific binding.

6. To determine the time-course of the association of ^{125}I-progesterone with its binding protein at different temperatures the above experiment was performed at four temperatures (4, 10, 25, and 37°C).

Calculation

1. The value of ^{125}I-progesterone bound specifically in (micro mole of ^{125}I-progesterone per mg protein) was calculated according to the following formula:

$$\text{The value of specifically bound } ^{125}\text{I - progesterone } (\mu\,\text{mol/mg}) = \frac{\text{specifically bound } ^{125}\text{I - progesterone in (molar)} \times \text{incubation volume in (litter)}}{\text{mg of protein in incubation medium}}$$

$$\text{specifically bound } ^{125}\text{I - progesterone (molar)} = \text{The percent of specific binding} \times \text{Total concentration of } ^{125}\text{I - progesterone in incubation medium (M)}$$

$$\text{The percent of specific binding} = \left(\frac{\text{Total binding (CPM) - Non specific binding (CPM)}}{\text{Total count (CPM) of } ^{125}\text{I - progesterne used in each tube}} \right) \times 100$$

2. The percent of specific binding was plotted against the time of incubation

2.2.6.2 THE THERMODYNAMIC OF ^{125}I-PROGESTERONE BINDING TO ITS RECEPTORS IN PRE MENOPAUSAL PATIENTS WITH UTERINE CANCER

Two hundred micro-liter of uterine homogenate (200µg protein) was incubated with $(6.369 \times 10^{-8}M)$ of ^{125}I-progesterone at 25°C for 16h. The final volume (1mL) was made up by adding the assay buffer (0.01M TED buffer pH 7.6). The steps 2,3,4 and 5 of the experiment (2.2.6.1) were carried out at four temperatures (4,10,25 and 37°C).

Calculation

1. The thermodynamic parameters of standard state were obtained from Van't Hoff plot, the values of the natural logarithm of equilibrium constant (affinity constant Ka) obtained at different temperatures were plotted against the reciprocal values absolute temperature in Kelvin (1/T), according to the following equation :

$$\ln Ka = \frac{\Delta S^\circ}{R} - \frac{\Delta H^\circ}{RT}$$

Where

ΔH° = the enthalpy change of the standard state.

ΔS° = the entropy change of the standard state.

R   = the gas constant (8.31441 JK^{-1}).

ΔH° value obtained from the slope of the linear relationship of the plot . The change in Gibbs free energy of the standard state (ΔG°) was obtained from the following equation:

$$\Delta G^\circ = -RT \ln Ka$$

While the standard state entropy change was obtained from :

$$\Delta S^\circ = \frac{\Delta H^\circ - \Delta G^\circ}{T}$$

2. The thermodynamic parameters of the transition state were obtained from Arrhenius plot of $\ln K_{+1}$ values against $1/T$ values, that gives a linear relationship according to the following equation:

$$\ln K_{+1} = \ln A - \left(\frac{Ea}{RT}\right)$$

Where A = Arrhenius constant .

The value of apparent energy of activation (Ea) of the binding reaction can be determined from the slope of the straight line . The enthalpy of transition state $\Delta H^{\bullet}$ obtained from

$$\Delta H^{\bullet} = Ea - RT$$

Transition state free energy change is calculated from the following equation:

$$\Delta G^{\bullet} = -RT \ln K_{+1} + RT \ln (KT/h)$$

Where K and h were Boltzmann and Plank's constants which equals $(1.38 \times 10^{-23}$ Jdeg$^{-1})$, $(0.662 \times 10^{-33}$ J S$^{-1})$ respectively.

The change in entropy of the transition state ΔS^* is calculated from the following relation:

$$\Delta S^{\bullet} = \frac{\Delta H^{\bullet} - \Delta G^{\bullet}}{T}$$

2.2.7 SPECTROSCOPIC STUDIES ON PROGESTERONE RECEPTORS

Absorption measurements were carried out by an Varian DMS 100 UV-visible spectrophotometer.

2.2.7.1 THE U.V SPECTRUM OF PROGESTERONE RECEPTORS IN EACH OF PREMENOPAUSAL PATIENTS WITH BENIGN UTERINE TUMOR, PRE AND POSTMENOPAUSAL PATIENTS WITH UTERINE CANCER

One hundred micro-liter (250μg protein) of cytosolic fraction was completed to 3mL with 0.01M TED buffer pH 7.8, then placed in the cuvette and the absorption spectrum was measured.

2.2.7.2 FACTORS AFFECTING THE ABSORPTION PROPERTIES OF PROGESTERONE RECEPTORS IN EACH OF PREMENOPAUSAL PATIENTS WITH BENIGN UTERINE TUMOR, PRE AND POST MENOPAUSAL PATIENTS WITH UTERINE CANCER

1- *pH effect on the u.v spectrum of progesterone receptors*:

One hundred micro-liter (250μg protein) of cytosolic fractions were completed to 3mL with 0.01M TED buffer at different pH (7.0, 7.4, 7.6, 7.8, 8.2, 8.6) then each of which were placed in the test cell and the TED buffer at adjusted pH was placed in the reference cell. Then the absorption spectrum was measured.

2- *Polarity effects on the uv spectrum of progesterone receptors*:

a- Effect of 10% ethanol on the progesterone receptor spectrum:

One hundred micro-liter (250μg protein) of cytosolic fractions were completed to 3mL with TED buffer contains 10% ethanol at different pH (7.6,

8.6) then each of which were placed in the test cell and the TED buffer at adjusted pH was placed in the reference cell. The absorption spectrum of each sample was measured.

b- Effect of 10% polyethylene glycol on the progesterone receptor spectrum:

One hundred micro-liter (250µg protein) of cytosolic fractions were completed to 3mL with TED buffer contains 10% PEG at pH 7.0, then placed in the cuvette and the absorption spectrum of each sample was measured.

c- Effect of urea on the progesterone receptor spectrum:

One hundred micro-liter (250µg protein) of cytosolic fractions were completed to 3mL with TED buffer contains 8mM urea at two different pH (7.0, 8.6), then each of which were placed in the test cell and the TED buffer at adjusted pH was placed in the reference cell. The absorption spectrum of each sample was measured.

2.2.7.3 PH TITRATION OF PROGESTERONE RECEPTORS IN EACH OF PRE MENOPAUSAL PATIENTS WITH BENIGN UTERINE TUMOR, PRE AND POST MENOPAUSAL PATIENTS WITH UTERINE CANCER

Cytosolic fractions (250µg protein in 100µl) were completed to 3mL with 0.01M TED buffer at different pH (7.8 - 8.6) then each of which were placed in the test cell and the TED buffer at adjusted pH was placed in the reference cell. The u.v absorption spectrum was measured in the area of (280- 290)nm.

2.2.7.4 SOLVENT-PERTURBATION ON PROGESTERONE RECEPTORS IN EACH OF PREMENOPAUSAL PATIENTS WITH BENIGN UTERINE TUMOR, PRE AND POSTMENOPAUSAL PATIENTS WITH UTERINE CANCER

The u.v spectrum of progesterone receptors were measured in the presence and absence of different perturbants:

1- 25% (w/v) polyethylene glycol at pH 7.0.

2- 8mM urea at pH 7.0.

3- 8mM urea and 0.03M KCL at pH 7.0.

One hundred micro-liter (250µg protein) of cytosolic fraction was completed to 3mL with TED buffer pH 7.0 with and without the addition of PEG. Then the absorption spectrum was measured in the presence and absence of PEG. The experiment was repeated with different perturbants.

CHAPTER THREE

Results

&

Discussion

3.1 BINDING STUDIES OF ^{125}I-PROGESTERONE WITH THEIR RECEPTORS IN BENIGN AND MALIGNANT UTERINE TUMOR

3.1.1 DETERMINATION OF PROGESTERONE LEVELS IN SERA OF UTERINE TUMOR PATIENTS

Serum progesterone levels were measured in pre and postmenopausal of uterine tumor patients matched with two groups of control subjects. Group I contained (8) premenopausal patients with endometrial cancer, Group II comprised of (6) postmenopausal patients with endometrial adenocarcinoma. Table(3-1) and figure(3-1) show the results obtained from this study. The level of serum progesterone in premenopausal patients was found to be 0.75ng/mL whereas that of control was found to be 0.93ng/mL. But in postmenopausal patients and the control, the level were found to be 0.046 and 0.038ng/mL respectively.

No significant increase and decrease of serum progesterone levels in these two groups of uterine cancer patients were obtained, this may be due to the similar phase of all specimens that have been used in this study (follicular phase). In this phase, the progesterone level is usually 0.9ng/mL (3nmol/L)[117].

Table (3-1): Serum progesterone levels (ng/mL) in pre and postmenopausal of uterine tumor patients. Details are described in section (2.2.3)

Group	No. of cases	Age (year)	Serum progesterone (ng/mL)
Premenopausal of uterine cancer	8	36.0 ± 5.2	0.75 ± 0.54
Control	16	38.0 ± 4.4	0.93 ± 0.62
Postmenopausal of uterine cancer patients	6	60.0 ± 1.9	0.046 ± 0.022
Control	3	62.0 ± 5.5	0.038 ± 0.017
Premenopausal of benign uterine tumor	14	38.0 ± 5.4	0.86 ± 0.51

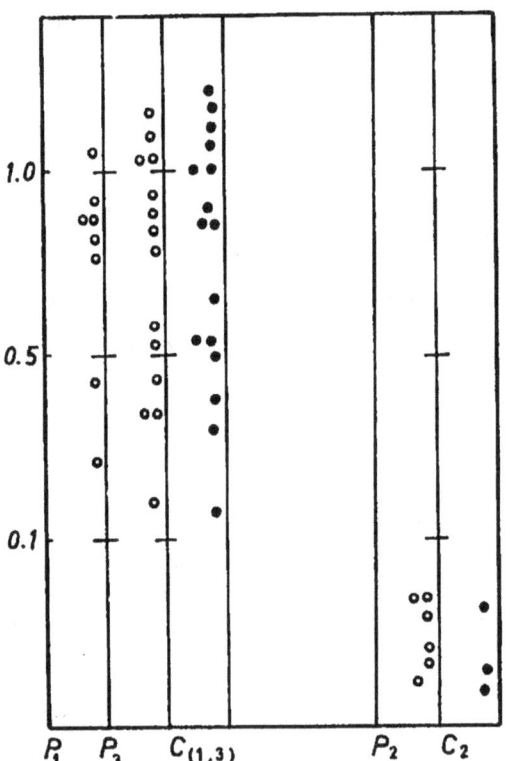

Figure(3-1) : Distribution of serum progesterone (ng/mL) in normal patients with benign uterine tumors, pre and post menopausal women with uterine cancer
P_1: Premenopausal patients with uterine cancer
P_2: Postmenopausal patients with uterine cancer
P_3: Premenopausal with benign tumor
C(1,3): Premenopausal normal women
C_2: Postmenopausal normal women
Details are described in section(2.2.3).

3.1.2 PRELIMINARY TEST OF ^{125}I-PROGESTERONE BINDING TO ITS RECEPTORS IN UTERINE TUMOR HOMOGENATE

Similar to section (2.1.3) three groups of uterine tumor patients were investigated to evaluate the presence of progesterone receptors after removal of the tumors surgically. Cytosolic progesterone receptors were detected through the incubation of (^{125}I-progesterone) with crude cytosol and the bound hormone was separated by dextran-coated charcoal method and then measured by gamma counter. The tumor was considered to have progesterone receptors, if it contained more than 5% of specific binding. The specific binding was found to be 31% in premenopausal of malignant uterine tumor patients, 29% in postmenopausal of malignant uterine tumor patients and 24.5% in benign uterine tumor patients table(3-2).

The data obtained, in this study revealed also that the tumors of uterine cancer patients have higher incidence of progesterone receptors than those of benign groups. The results are in consistent with those reported previously[118]. In addition, and as indicated from our results, the tumors of premenopausal uterine cancer patients include higher incidence of progesterone receptors than those of postmenopausal patients.

Total progesterone receptors are usually high in the late proliferative phase (about 12000sites/cell) and significantly lower in the late secretory phase[119,120]. During the proliferative phase, progesterone receptors are increased mainly in the cytoplasm. In the early luteal phase, progesterone receptors decreased in the cytosol, whereas they remained high in the nuclei. Progesterone receptors are at lowest level in cytosol and nuclei in the late secretory phase. The changes of total progesterone receptor sites and of its respective subcellular distributions seem to depend upon the plasma levels of the hormone and follow the same pattern and effect relationships as those demonstrated experimentally in laboratory animals. Thus accordingly all the specimens that used in this study were in the follicular phase.

Table(3-2): Incidence of progesterone receptors in benign and malignant uterine tumors. Details are described in section(2.2.4)

Group	No. of cases	Cytosolic progesterone receptors
Premenopausal of uterine cancer patients	8	31%
Postmenopausal of uterine cancer patients	6	29%
Premenopausal patients of benign uterine tumors	14	24.5%

3.2 RADIO RECEPTOR ASSAY STUDIES OF ^{125}I-PROGESTERONE BINDING TO ITS RECEPTORS IN BENIGN AND MALIGNANT UTERINE TUMOR

3.2.1 EFFECT OF ^{125}I-PROGESTERONE CONCENTRATION ON THE BINDING WITH ITS RECEPTORS IN UTERINE TUMOR HOMOGENATE

One of the criteria of the concept of hormone receptors is its saturability. To estimate the suitable concentration of ^{125}I-progesterone. Cytosolic samples (200µg protein in 200µL) were incubated with increasing concentrations of ^{125}I-progesterone for 16h at 4°C. The results revealed that the progesterone binding by cytosolic fractions were increased with increasing amount of ^{125}I-progesterone added. Figure (3-2) is a representative of ^{125}I-progesterone binding curve for a cytosolic fractions. Accordingly, in all the subsequent experiments, 20ng per 40µL of labeled progesterone was used, since it gives highest binding.

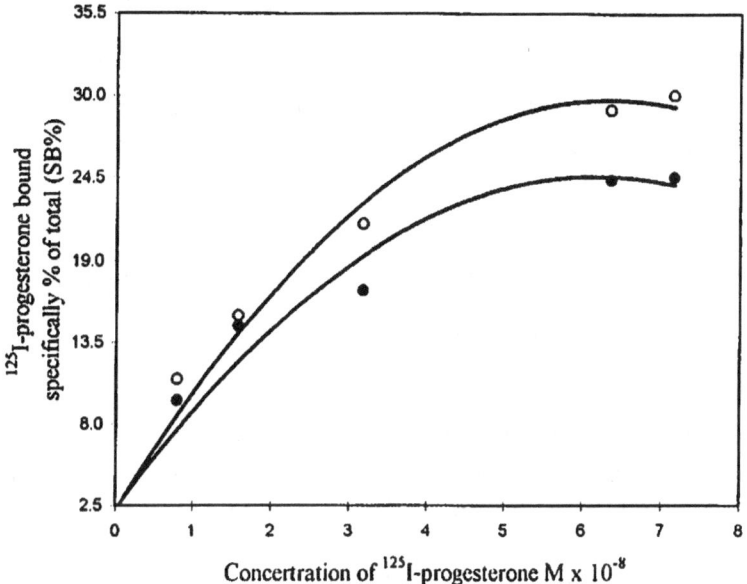

Figure(3-2): Effect of the concentration of ^{125}I-progesterone on the binding with,
    • **Benign uterine tumor homogenate**
    o **Malignant uterine tumor homogenate**
   Details are described in section (2.2.5.1).

3.2.2 EFFECT OF PROGESTERONE RECEPTOR CONCENTRATION ON THE BINDING IN UTERINE TUMOR HOMOGENATE

To determine whether the specific binding was proportional to the amount of protein of the receptors used, increasing amounts of cytosol were incubated with ^{125}I-progesterone or with nonradioactive progesterone, according to the details in section(2.2.5.2). The linear relationship observed from figure(3-3) indicates that the cytosol preparations were homogenous in terms of receptor distribution. The specific binding was increased when the amount of receptor protein in the incubation mixture was increased. In all the subsequent experiments, 200µg of receptor protein in the incubation mixture was used, according to the results obtained in this experiment.

3.2.3 EFFECT OF TEMPERATURE ON THE BINDING OF ^{125}I-PROGESTERONE TO ITS RECEPTORS IN UTERINE TUMOR HOMOGENATE

Temperature dependency of the association of radioactive progesterone to its cytosolic receptors was investigated. Cytosol fractions of benign and malignant uterine tumors were incubated for 16h at different temperatures (4-40°C). Figure(3-4) revealed that the specific binding of ^{125}I-progesterone to its cytosolic receptors was increased when the temperature was raised from 4 to 25°C and a maximal binding was obtained at 25°C. These results are in consistent with those reported by other investigators and a decrease in the specific binding at 37°C was probably due to denaturation of receptor molecules[121]. According to these results, 25°C was used in all the subsequent experiments of the radio receptor assay studies.

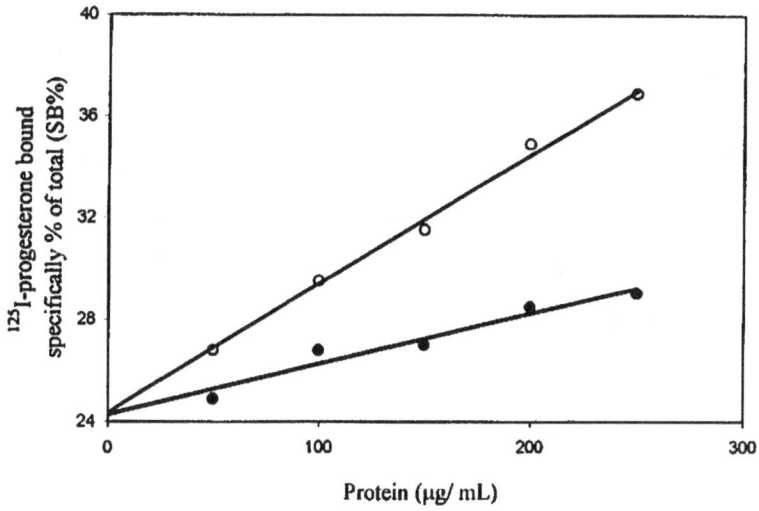

Figure(3-3): Influence of receptor concentration on the binding of
¹²⁵I-progesterone with,

• **Benign uterine tumor homogenate**
o **Malignant uterine tumor homogenate**
Details are described in section (2.2.5.2).

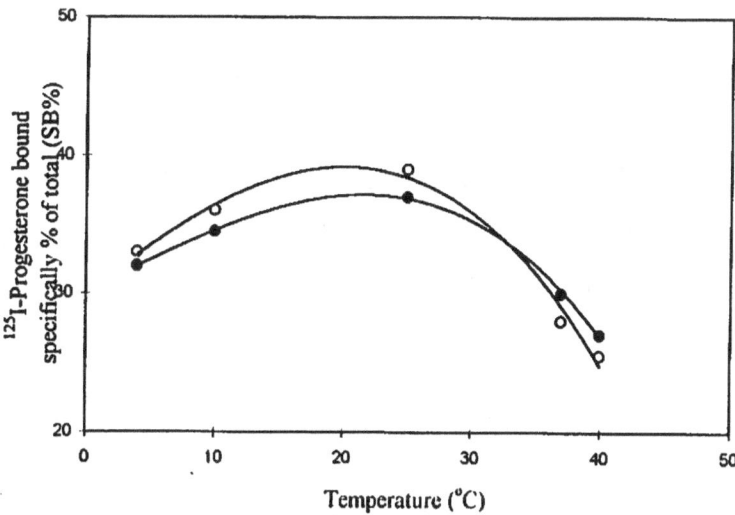

Figure(3-4):Effect of the temperature on the binding of
¹²⁵I-progesterone with,

• **Benign uterine tumor homogenate**
o **Malignant uterine tumor homogenate**
Details are described in section (2.2.5.3).

3.2.4 TIME-COURSE OF RECEPTOR BINDING IN UTERINE TUMOR HOMOGENATE

The time-course of the binding of progesterone to its receptors was investigated by incubating cytosol fractions of benign and malignant uterine tumors for the time indicated at 25°C with or without nonradioactive progesterone. Figure(3-5) shows the results of this analysis. It seemed, that the specific binding of ^{125}I-progesterone to its receptors was maximal at 16h. In all other subsequent experiments that 16h and 25°C were implemented.

3.2.5 EFFECT OF PH ON THE BINDING OF ^{125}I-PROGESTERONE TO ITS RECEPTORS IN UTERINE TUMOR HOMOGENATE

The effect of pH on the specific binding of ^{125}I-progesterone to its receptors was investigated. Figure(3-6) shows that the optimal binding of radioactive progesterone to its receptors was achieved by incubating the cytosol (receptors) with ^{125}I-progesterone for 16h at pH 7.6 and 25°C. According to the results obtained in this analysis, the pH of the buffer used in all subsequent experiments was adjusted to 7.6.

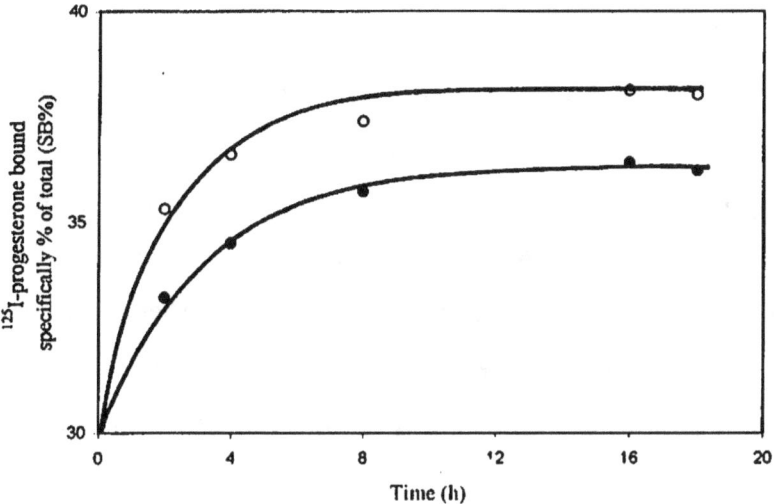

Figure(3-5):Time-course of ^{125}I-progesterone binding with,
- Benign uterine tumor homogenate
- o Malignant uterine tumor homogenate

Details are described in section (2.2.5.4).

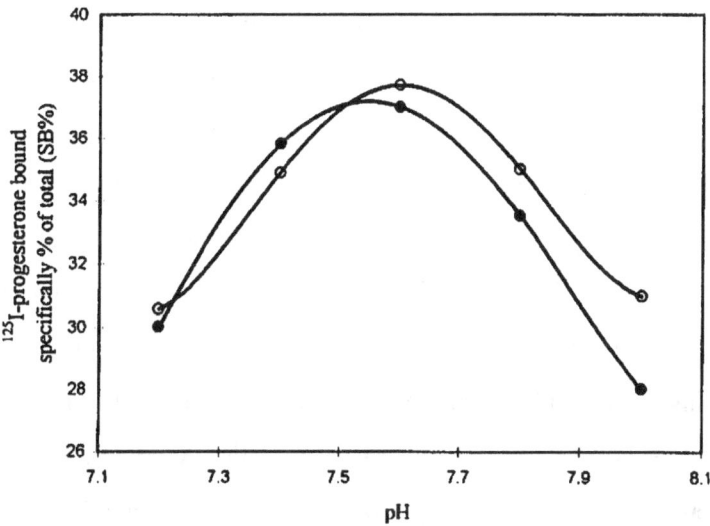

Figure(3-6): pH dependency of ^{125}I-progesterone with,
- Benign uterine tumor homogenate
- o Malignant uterine tumor homogenate

Details are described in section (2.2.5.5).

3.2.6 STABILITY OF ^{125}I-PROGESTERONE RECEPTOR COMPLEX

The influence of temperature on the stability of progesterone-receptor complex as a function of time was studied. The complex was reincubated at three temperatures (0, 25, 45°C) and at a certain time intervals the remaining bound hormone was estimated. As seen in figure(3-7), the dissociation of progesterone-receptor complex was relatively increased when the temperature was also increased. At 25°C progesterone-receptor complex was found to be more stable than at 45°C. This finding is consistent with the same observations found in a number of studies[119,122].

3.2.7 COMPETITIVE EFFECT OF DIFFERENT CONCENTRATIONS OF UNLABELED ESTRIOL AND PROGESTERONE ON THE BINDING OF ^{125}I-PROGESTERONE TO ITS RECEPTORS IN UTERINE TUMOR HOMOGENATE

The specificity of cytoplasmic ^{125}I-progesterone binding sites (receptors) of malignant uterine tumor homogenate was demonstrated by a decrease in receptor bound radioactivity after incubating the cytosol with increasing amounts of unlabeled steroids hormone. The binding of ^{125}I-progesterone to its receptors was effectively inhibited by unlabeled progesterone but not estriol.

Thus, the receptors binding indicates that the most potent steroid in the target tissue is progesterone itself, since the maximal competition was obtained at 200 fold higher of unlabeled progesterone than the concentration of labeled progesterone. Several studies have illustrated the competitive effect of unlabeled steroids on the binding of ^{125}I-progesterone to its receptors in uterine tumor tissue[123]. Figure(3-8) shows the effect of different concentrations of unlabeled estriol and progesterone and the shape of the curve ascertained the relationship between labeled and unlabeled hormones.

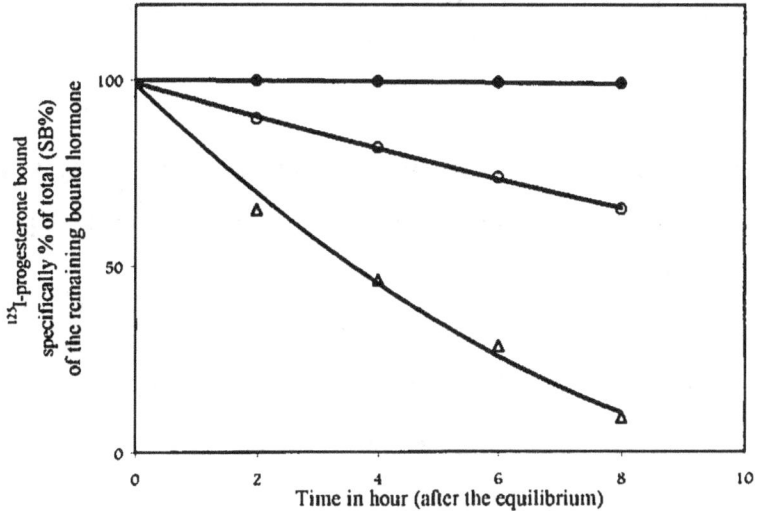

Figure(3-7): Stability of (¹²⁵I-progesterone receptor) complex of malignant uterine tumor homogenate at three different temperatures,
(•) 0°C   (o) 25°C   (Δ) 45°C
Details are described in section (2.2.5.6).

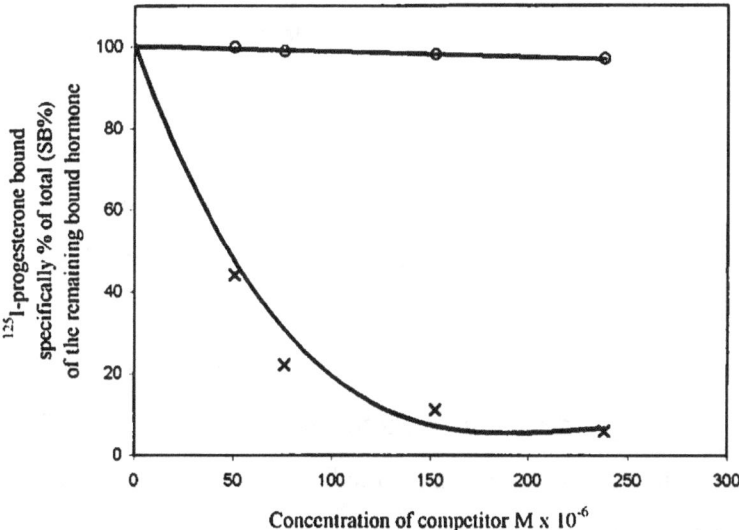

Figure(3-8): Binding of malignant uterine tumor homogenate receptors with ¹²⁵I-progesterone in the presence of different concentrations of unlabeled
(×) Progesterone   (o) Estriol
Details are described in section (2.2.5.7).

3.2.8 EFFECT OF DIFFERENT HALIDES ON THE BINDING OF ^{125}I-PROGESTERONE TO ITS RECEPTORS

Different halides of sodium were investigated to study their action on the binding of progesterone with its receptors in uterine cytosol. The sodium halides in the incubation mixture induced-activation of the percent of specific binding according to the following sequence:

NaCl >NaI > NaF

This frequency of binding indicate that activation causes subtle conformational changes in the progesterone receptor. These results are shown in figure(3-9), which illustrates somewhat the relationship between the halides and the progesterone receptors.

3.2.9 EFFECT OF DIVALENT CATIONS ON THE BINDING OF ^{125}I-PROGESTERONE WITH RECEPTORS OF UTERINE TUMOR HOMOGENATE

This study was devised to explore the possible effect of certain metal ions on the binding of progesterone with its receptors in uterine tumor homogenate. Progesterone-binding process was found to be sensitive to the presence of certain metal ions. The cations seemed to interfere directly with SH groups at the progesterone binding site. The presence of Cu(II) ions at (25mM) are capable of inhibiting progesterone binding.

These results support the view that SH groups are involved in the processes leading to association of progesterone with its cytosolic binding protein. The influence of metal ions on the binding of progesterone to cytosolic fractions are summarized in figure(3-10).

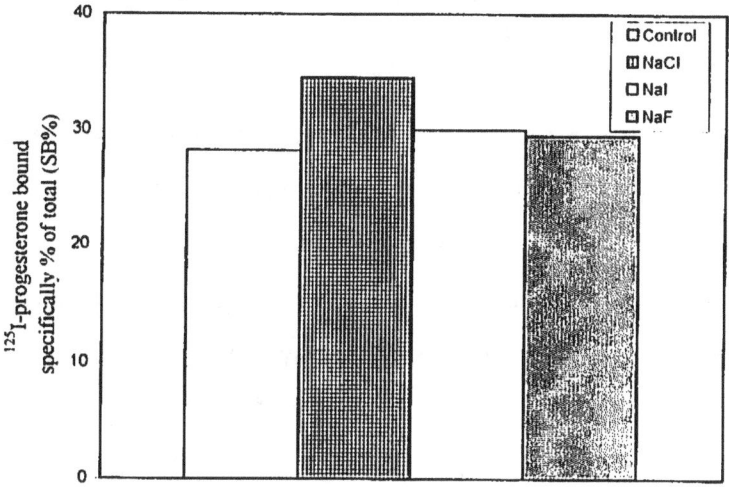

Figure (3-9): Effect of (0.01 M) concentration of NaCl, NaI, NaF on the extent of ^{125}I-progesterone binding to its receptor in uterine tumor homogenate. Details are described in section (2.2.5.8).

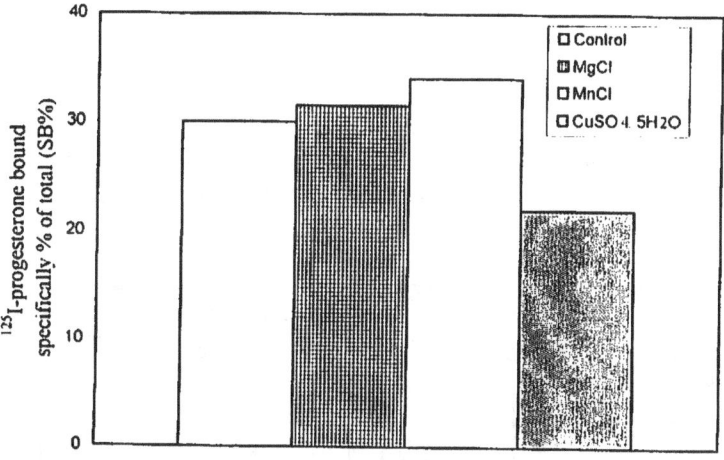

Figure (3-10): Effect of different solts (25 mM) on the extent of ^{125}I-progesterone binding to its receptor in uterine tumor homogenate. Details are described in section (2.2.5.9).

3.2.10 DETERMINATION OF THE CONCENTRATION AND AFFINITY CONSTANT OF PROGESTERONE RECEPTORS

Cytosolic progesterone receptor concentration and the affinity constant of the binding have been measured in uterine tumors that show specific binding in the preliminary test. The experiment was carried out at the optimal conditions, which were obtained in previous experiments. Following a 16h incubation at 25°C, dextran-coated charcoal was used to assay the post incubation of cytosols for total, non-specific and specific binding of ^{125}I-progesterone.

Scatchard plot analysis gave straight line as shown in figure (3-11). The results obtained indicate the presence of only one species of receptor site, or more but with the same affinity and number of binding sites. The results are summarized in table(3-3). Cytosolic progesterone receptor binding capacities of postmenopausal uterine cancer patients were 187.5pmol/mg protein while of premenopausal patients were 202.5pmol/mg protein and of premenopausal patients with benign uterine tumors were 82.5pmol/mg protein.

Table (3-3): Concentration and affinity constant of cytosolic progesterone receptors in three groups of uterine tumor patients. Details are described in section (2.2.5.10)

Group	No. of cases	Age (year)	Binding capacity pmol/mg	Ka $M^{-1} \times 10^7$
Premenopausal of uterine cancer patients	8	36.0 ± 5.2	202.5	2.73
Postmenopausal of uterine cancer patients	6	60.0 ± 1.9	187.5	2.54
Premenopausal patients of benign uterine tumors	14	38.0 ± 5.4	82.5	2.30

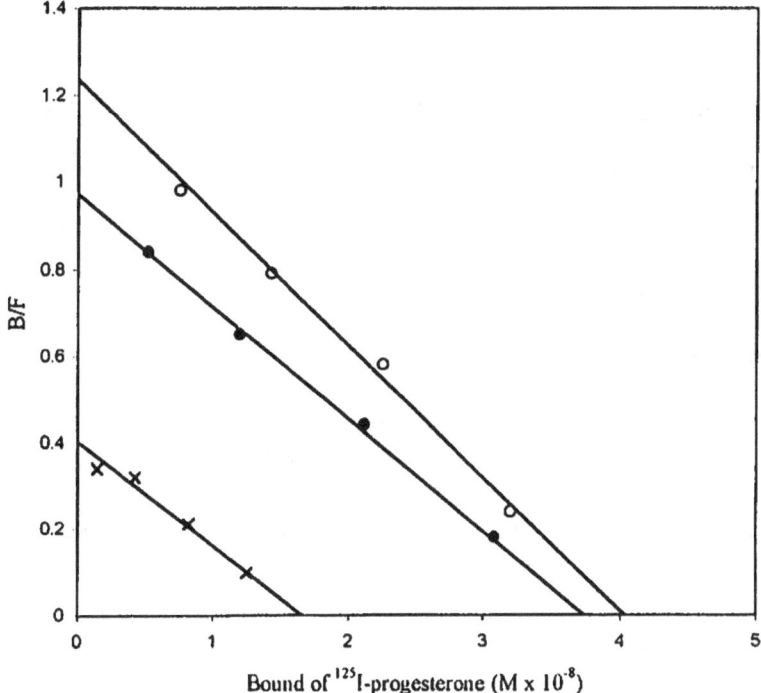

Figure(3-11): Scatchard plot of ^{125}I-progesterone binding with its receptors in uterine tumors homogenate of the three groups studied ,

(o)Malignant tumors of premenopausal patients
(•)Malignant tumors of postmenopausal patients
(x) Benign tumors of premenopausal patients

Details are described in section (2.2.5.10).

3.3 THE KINETIC AND THE THERMODYNAMIC STUDIES

3.3.1 KINETICS OF THE ^{125}I-PROGESTERONE BINDING TO ITS RECEPTORS IN PREMENOPAUSAL PATIENTS WITH MALIGNANT UTERINE TUMOR

Figure(3-12) shows the time-course of the formation of ^{125}I-progesterone-receptor complex at four different temperatures (4, 10, 25 and 37°C) in premenopausal patients with uterine cancer. The concentration of ^{125}I-progesterone-receptor complex that formed after time (t) was calculated from the following equation:

$$\begin{bmatrix} ^{125}\text{I} - \text{progesterone} - \text{receptor} \\ \text{in Molar formed} \\ \text{after time (t)} \end{bmatrix} = \frac{\text{Count (CPM) of } ^{125}\text{I} - \text{progesterone specifically bound after time (t)}}{\text{Total count (CPM) of } ^{125}\text{I} - \text{progesterone used in the incubation}} \times \begin{array}{c} \text{Concentration of the} \\ ^{125}\text{I} - \text{progesterone in} \\ \text{the incubation} \\ \text{medium (Molar)} \end{array}$$

The result of time-course pattern at different temperatures revealed that the binding of ^{125}I-progesterone to its receptors in uterine tumor homogenate is a temperature and time dependent process with a maximum binding occurs at 25°C and 16h.

3.3.2 DETERMINATION OF KINETIC PARAMETERS OF ^{125}I-PROGESTERONE BINDING TO ITS RECEPTORS IN PREMENOPAUSAL PATIENTS WITH MALIGNANT UTERINE TUMORS

The time-course of ^{125}I-progesterone to its receptors in uterine tumor homogenate, was carried out to describe the kinetic parameters of the binding. The simplest proposed model representing the interaction of ^{125}I-progesterone with its receptors could be expressed by the following equation :

$$^{125}\text{I} - \text{progesterone} + \underset{(\text{receptor})}{\text{R}} \underset{K_{-1}}{\overset{K_{+1}}{\rightleftarrows}} {}^{125}\text{I} - \text{progesterone} - \text{R}$$

Where K_{+1} is the rate of the association of ^{125}I-progesterone with its receptors and K_{-1} represents the rate of the reverse reaction of the dissociation of the complex formed under the same conditions. At equilibrium:

$$Ka = \frac{[^{125}I\text{-}progesterone\text{-}R]}{[^{125}I\text{-}progesterone][R]} \tag{1}$$

$$Kd = \frac{[^{125}I\text{-}progesterone][R]}{[^{125}I\text{-}progesterone\text{-}R]} \tag{2}$$

Thus,

$$Ka = \frac{1}{Kd} = \frac{K_{+1}}{K_{-1}} \tag{3}$$

Where Ka is the equilibrium constant of the association (affinity constant) and Kd is the equilibrium constant of dissociation of ^{125}I-progesterone-R complex.

The values of Ka and maximal binding capacity (Bmax) were calculated from Scatchard plot at four different temperatures (table(3-4) and figure(3-13)).

Table (3-4): The kinetic parameters of ^{125}I-progesterone binding to its receptors in uterine tumor homogenate. Details are described in section (2.2.6.1).

Temp (°C)	Binding capacity pmol/mg protein	Kd=K$_{-1}$/K$_{+1}$ x10^{-8} (M)	Ka=K$_{+1}$/K$_{-1}$ x 10^{7} (M^{-1})
4	170.0	4.78	2.09
10	180.0	4.42	2.26
25	202.5	3.66	2.73
37	175.0	7.40	1.35

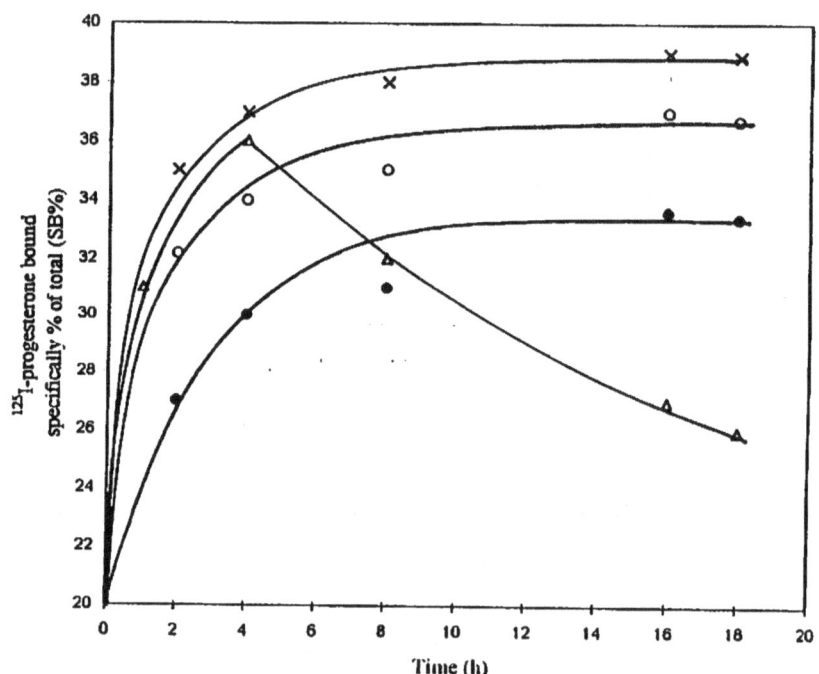

Figure(3-12): Time-course of cytosolic progestrone recepter at different temperatures,
(•)4°C   (o)10°C   (×) 25°C   (Δ) 37°C
Details are described in section (2.2.6.1).

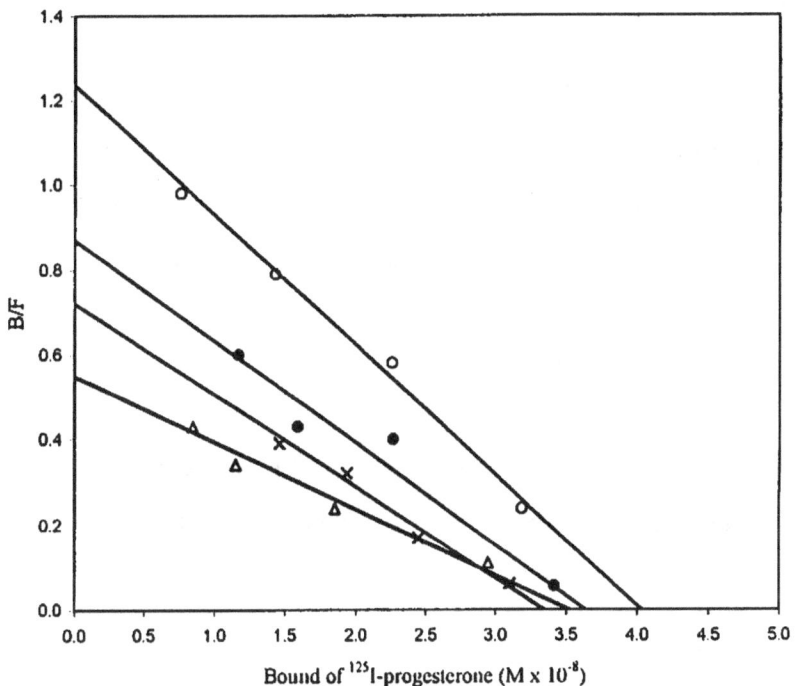

Figure(3-13): Scatchard plot of ^{125}I-progesterone binding with its receptors in premenopausal patients of uterine cancer at different four temperatures,

(×) 4°C   (•) 10°C   (o) 25°C   (Δ) 37°C

Details are described in section (2.2.6.1).

Different equations are used for the determination of the association rate constant (K_{+1}) of ^{125}I-progesterone with its receptors at four temperatures and then for verification of the order of reaction. These include the followings:

$$1) \ \ln \overline{x} \left[\left\{ L_o - \left(\frac{\overline{x}x}{R_o} \right) \right\} \bigg/ L_o (\overline{x} - x) \right] = K_{+1} \ t \left[\frac{\left(L_o R_o - \overline{x}^2 \right)}{\overline{x}} \right] \tag{4}$$

Where K_{+1} is the kinetic association constant in $M^{-1}S^{-1}$, L_o is the total ^{125}I-progesterone concentration in molar, R_o the total concentration of the progesterone-binding sites, $\overline{x}$ the concentration of ^{125}I-progesterone- receptor complex formed at equilibrium and x is the concentration of the complex formed after time (t).

$$2) \ \frac{1}{S_o - Q_o} \cdot \ln \left(\frac{S_o - B(t)}{Q_o - B(t)} \right) = K_{+1} \cdot t + \frac{1}{S_o - Q_o} \cdot \ln \left(\frac{S_o}{Q_o} \right) \tag{5}$$

Where S_o is the total concentration of progesterone, Q_o is the total concentration of receptor binding sites, $B(t)$ is the concentration of specifically bound progesterone at time (t) and K_{+1} is the association rate constant in $M^{-1}S^{-1}$.

$$3) \ \ln \frac{(HR)_e}{(HR)_e - (HR)_t} = t \cdot K_{obs} \tag{6}$$

$$K_{obs} = K_{+1} \ \frac{(H)_T \times (R)_T}{(HR)_e} \tag{7}$$

Where $(HR)_e$ the concentration of progesterone-receptor complex at equilibrium, $(HR)_t$ the concentration of progesterone-receptor complex formed after time (t), K_{obs} the observed value of the association rate constant, K_{+1} the association rate constant in min^{-1}, $(H)_T$ the total progesterone concentration and $(R)_T$ the total receptor concentration.

The time-course data obtained from figure (3-12) could be used to confirm the suggested second order kinetics for the association of ^{125}I-progesterone with its receptors in uterine tumor homogenate, accordingly, the following equation that was used[124] :

$$\frac{1}{S_o - Q_o} \cdot \ln\left(\frac{S_o - B(t)}{Q_o - B(t)}\right) = K_{+1} \cdot t + \frac{1}{S_o - Q_o} \cdot \ln\left(\frac{S_o}{Q_o}\right)$$

For this purpose, $\ln((S_o-B(t))/(Q_o-B(t)))$ was plotted against time (t) as shown in figure(3-14). The solid lines obtained were fitted with a second-order reaction of the data taken at 4°C, 10°C, 25°C and 37°C. K_{+1} was determined from the slope of the plot.

Since in some cases of our work the percent of the specific binding was small (less than 10% of the total concentration of ^{125}I-progesterone used in incubation medium) and most of the ^{125}I-progesterone remained free and only a small fraction of S_o was bound even at equilibrium (Pseudo–first order conditions), so that the following equation[125] could be used in order to fit the data of the first-order kinetics:

$$\ln\frac{(HR)_e}{(HR)_e - (HR)_t} = t \cdot K_{obs}$$

Figure(3-15) shows that the plotting $\ln\dfrac{(HR)_e}{(HR)_e - (HR)_t}$ against time (t) gives a straight line with a slope equal to the observed value of first-order rate constant, and the rate constant K_{+1} was calculated from equation[126]:

$$\ln K_{+1} = \ln A - Ea/RT$$

The half life time of association $(t_{1/2})_{ass.}$, which represents the time needed for the formation of half amount of the complex at equilibrium, was determined from the concentration of the complex at equilibrium and the time-course curve. While the half life time of dissociation $(t_{1/2})_{diss.}$ was determined from:

$$\left(t_{1/2}\right)_{diss} = \ln\frac{2}{K_{-1}} = \frac{0.693}{K_{-1}}$$

The Ka values were also obtained from equation(3). Figure(3-14) represent the kinetics of complex formation between ^{125}I-progesterone and its receptors in uterine tumor homogenate at different temperatures.

The results revealed that the association rate constant K_{+1} at 25°C was higher than that of at 4, 10 and 37°C as shown in table(3-5). The values of K_{-1} were obtained also from the values of Ka which have been estimated at the four temperature investigated. The K_{-1} was determined from the equation(3). Table(3-5) showed that K_{-1} increased with the elevation of temperature. Thus when the reaction temperature was increased from 4°C to 25°C, the value of the association constant increased approximately 1.5 folds while that of the dissociation constant was increased 1.2 folds.

Table(3-5): **The effect of temperature on the kinetic parameters of progesterone binding to its receptors in uterine tumor homogenate. Details are described in section (2.2.6.1)**

Temp. (°C)	K_{+1} (M.min)$^{-1}$ $\times 10^3$	K_{-1} (min)$^{-1}$ $\times 10^{-4}$	Ka (M)$^{-1}$ $\times 10^7$	$(t_{1/2})_{ass}$ min	$(t_{1/2})_{diss}$ min
4	2.4	1.148	2.09	480	0.603
10	2.6	1.150	2.26	480	0.603
25	3.7	1.355	2.73	480	0.511
37	3.6	2.667	1.35	120	0.259

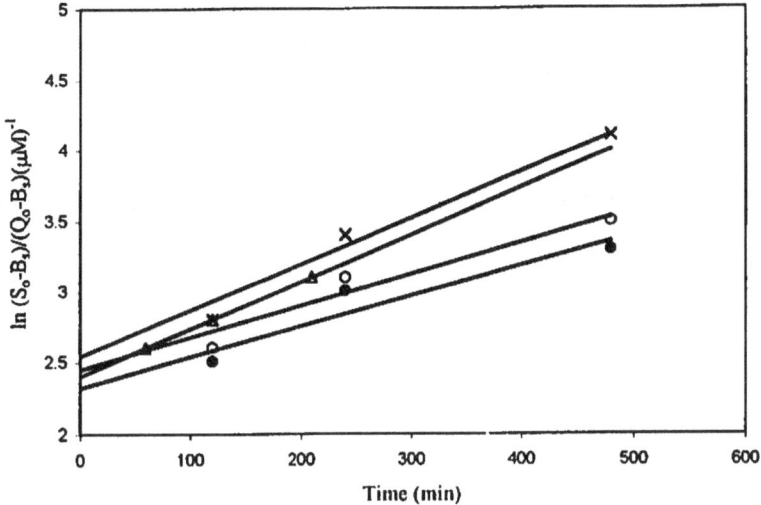

Figure(3-14): Kinetics of [125]**I-progesterone binding with the uterine cytoplasmic receptor,**

(•) 4°C   (o) 10°C   (×) 25°C   (Δ) 37°C

Details are described in section (2.2.6.1).

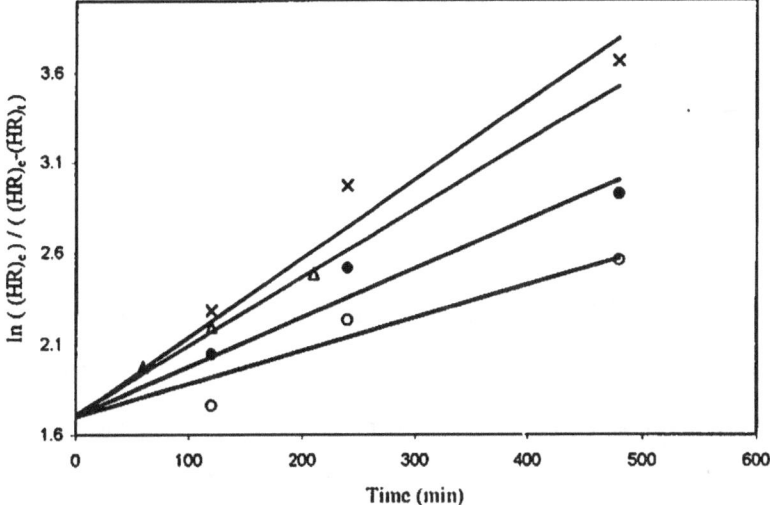

Figure(3-15) Kinetics of [125]**I-progesterone binding with the uterine cytoplasmic receptor,**

(o) 4°C   (•) 10°C   (×) 25°C   (Δ) 37°C

Details are described in section (2.2.6.1).

3.3.3 THE THERMODYNAMICS OF THE BINDING OF ^{125}I-PROGESTERONE TO ITS RECEPTORS IN PREMENOPAUSAL PATIENTS WITH MALIGNANT UTERINE TUMORS

1- Thermodynamic parameters of standard state

Figure(3-16) represent the dependence of the equilibrium binding constant (affinity constant) for the binding of ^{125}I-progesterone to its receptors in uterine tumor homogenate on the temperature (Van't Hoff plot).

The results indicated that $\Delta H°$ in general had small values and their positive sign ascertain that the reactions were nearly endothermic.

The negative values of $\Delta G°$ reflects the stability of the complex hence, the high affinity of the reactants. The high negative values of $\Delta G°$ for the binding reaction is controlled by high positive $\Delta S°$ as shown in table(3-6). So our system is characterized by the sole contribution of $\Delta S°$ to the stability of the complex formed, while $\Delta H°$ has little or no effect[127]. A high values of positive $\Delta S°$ suggest that the reaction spontaneity was entropically driven. Entropy was the driven force for the occurrence of the reaction (binding). This indicate that the hydrophobic interactions played an important role in stabilizing the complex[128].

The small positive $\Delta H°$ may indicate a favorable interaction between groups within both ^{125}I-progesterone and its receptors. These include the non-covalent interactions which are fundamentally electrostatic in nature such as charge-charge interactions which occur in both ^{125}I-progesterone and its receptors in uterine tumor homogenate, other types of interactions include charge-dipole, dipole-dipole, charge-induced dipole, dipole-induced dipole and hydrogen bond. The sum of these types of interactions can yield some stabilization to the folded structure of the complex. So the negative values of

ΔG° showed that the overall reaction was energetically favorable in the direction of complex formation.

Table(3-6): Thermodynamic parameters at standard state of progesterone binding to its receptor in uterine tumor homogenate. Details are described in section(2.2.6.2)

Temp. (°C)	ΔH° (KJ /mol)	ΔG° (KJ /mol)	ΔS° (J /mol.K)
4	9.26	-38.85	+173.588
10	9.26	-39.86	+173.477
25	9.26	-42.44	+173.403
37	9.26	-42.34	+166.371

2- Thermodynamic parameters of transition state

According to the transition state theory, the interaction of two proteins leads to the formation of an activated complex (transition state), then the formation of the final product:

$$^{125}I - progesterone + R \ \rightarrow \ [^{125}I - progesterone - R] \ \rightarrow \ ^{125}I - progesterone - R$$
$$\text{an activated complex}$$
$$\left(\text{transition state}\right)$$

The transition-state thermodynamic parameters ΔH^*, ΔG^* and ΔS^* could be determined from Arrhenius equation and the kinetics constant.

Figure(3-17) shows the dependent of the association rate for the binding of ^{125}I-progesterone to its receptors in uterine tumor homogenate on temperature (Arrhenius plot).

Table(3-7) the high positive values of ΔG^* indicated that the formation of an activated [^{125}I-progesterone–R] complex was a non spontaneous process and required a lot of energy (equal to Ea) to overcome the transition state energy barrier and giving the final product, whereas the high negative ΔS^* revealed that the activated complex had a more order structure than the reactants.

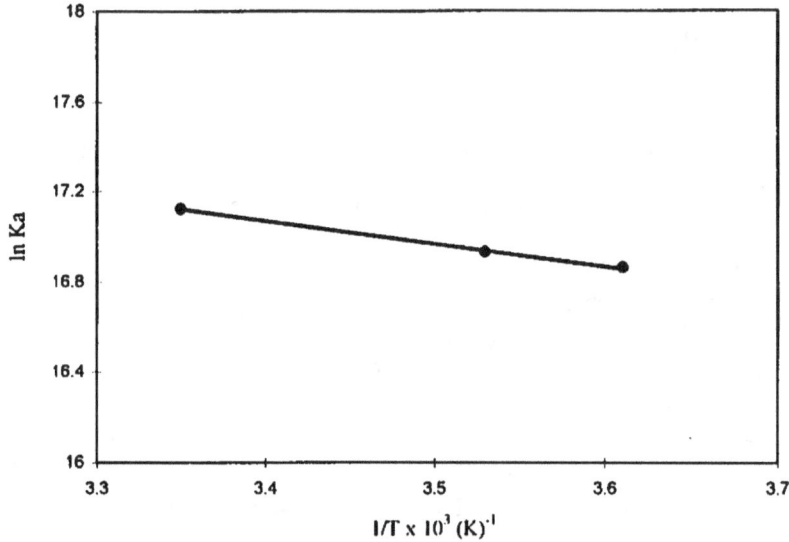

Figure (3-16): Van't Hoff plot for the ^{125}I-progesterone binding to its receptors in uterine tumor homogenate. Details are described in section (2.2.6.2)

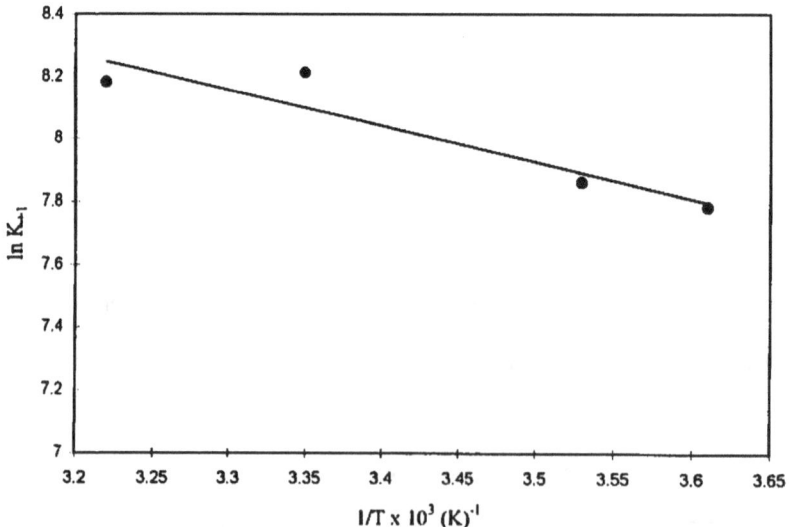

Figure (3-17): Arrhenius plot for the ^{125}I-progesterone binding to its receptors in uterine tumor homogenate. Details are described in section (2.2.6.2)

The positive values of ΔG^* is mainly attributed to the decreased in entropy of the transition state ($\Delta S^* < 0$). In addition the positive value of ΔH^* shows that the heat content of the activated complex is more than that of isolated species[129].

The deviation in the thermodynamic parameters at 37°C for the standard and transition states of the binding reaction in uterine tumor homogenate was higher than the other temperatures investigated, this could be attributed to the elevated temperature which affect the protein structure.

Determination of the thermodynamic parameters of the binding-reaction using the equilibrium data, gave an overall idea about the nature of forces that control the complex formation. From these results, a thermodynamic model describing the complex formation was suggested. The model is illustrated in figure(3-18).

This model proposes that the formation of the (^{125}I-progesterone- receptor) complex undergo three thermodynamic states. The thermodynamic state A represents the initial energy level of ^{125}I-progesterone and receptor (R). In the thermodynamic state B, the two species bind to form the activated complex (^{125}I-progesterone-R). The last thermodynamic state C, represents the fully interacting ^{125}I-progesterone-R complex. In step 1 of the reaction, the binding of ^{125}I-progesterone to its receptors was associated with positive ΔG^* value. This indicates that the initial step of the reaction requires input of energy for the system. The negative entropy change ΔS^* for this step of the reaction reflects the change of the ^{125}I-progesterone-R transition complex to a more ordered structure. In step 2, the activated complex participates in further interactions, giving the fully interacting complex (^{125}I-progesterone-R). It is proposed that the formation of a protein-ligand complex, occurs in two steps. The first is the stabilization of the complex by hydrophobic interactions and the second is the stabilization by short range interactions, such as electrostatic interactions, hydrogen bonding and Van der Waals interactions[130]. Hydrophobic interactions contribute to the complex stability via high positive entropy changes ($\Delta S > 0$), while electrostatic interactions, hydrogen bonding and Van der Waals interactions contribute to the stability of the complex via negative entropy change ($\Delta S^* < 0$) [130,131].

The thermodynamic data from our study indicate that the binding of ^{125}I-progesterone to its receptors are entropy driven and come in agreement with the concept that hydrophobic interactions play an important role in ^{125}I-progesterone-R interactions.

Table (3-7): Thermodynamic parameters at transition state of progesterone binding to its receptors in uterine tumor homogenate. Details are described in section (2.2.6.2).

Temp. (°C)	Ea (KJ/mol)	ΔH* (KJ/mol)	ΔG* (KJ/mol)	ΔS* (J/mol.K)
4	12.753	10.45	40.3	-107.703
10	12.753	10.40	41.1	-108.423
25	12.753	10.27	43.0	-109.777
37	12.753	10.17	44.4	-110.366

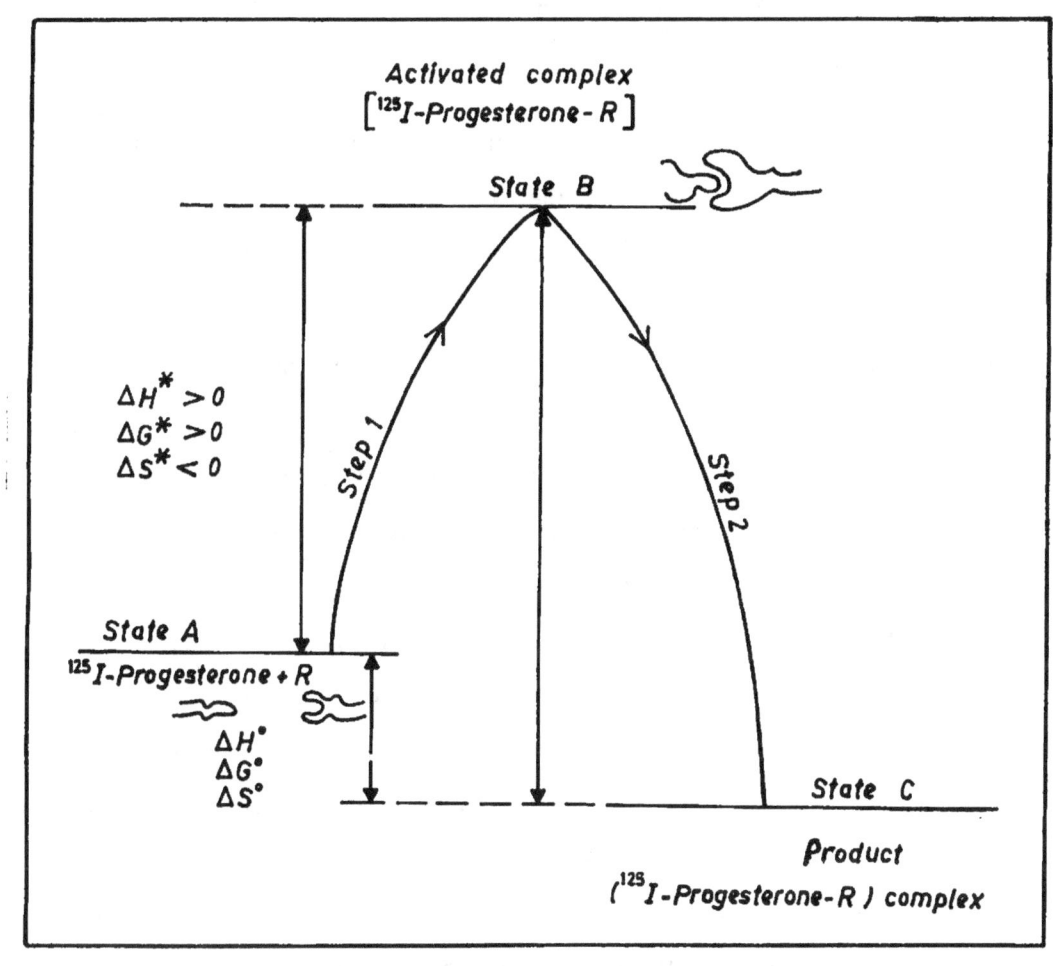

Reaction coordinate

Figure(3-18) :General energy diagram and thermodynamic model applied
to the complex formation between ^{125}I-progesterone and its
cytoplasmic receptor protein in uterine tumor homogenate

3.4 SPECTROSCOPIC STUDIES ON PROGESTERONE RECEPTORS

3.4.1 THE U.V SPECTRUM OF PROGESTERONE RECEPTORS IN EACH OF PREMENOPAUSAL PATIENTS WITH BENIGN UTERINE TUMOR PRE AND POSTMENOPAUSAL PATIENTS WITH UTERINE CANCER

Figure (3-19) illustrates the u.v spectrum of progesterone receptors at pH 7.8. The spectrum shows that the λ_{max} for progesterone receptors in premenopausal patients with benign uterine tumor is consisted of three peaks; at 286.9nm, 221.3nm and 203.8nm, in premenopausal patients with uterine cancer the λ_{max} is consisted of two peaks; at 272.6nm and 218.5nm and in postmenopausal patients with uterine cancer, the λ_{max} is consisted of two peaks; at 280.5nm and 219.9nm. As a result, each progesterone receptors have a characteristic spectrum and can be identified by their peaks. For premenopausal patients with benign uterine tumor, the peaks at 286.9, 221.3 and 203.8nm are assigned to tryptophan, tyrosine and phenylalanine respectively. For premenopausal patients with uterine cancer, the two peaks; at 272.6 and 218.5nm are also assigned to tyrosine and tryptophan respectively. For postmenopausal patients with uterine cancer, the two peaks; at 280.5 and 219.9nm are assigned to tryptophan only. It seems that in premenopausal patients with benign and malignant uterine tumor, tryptophan is located in a way, that part of it, is on the surface of the protein molecule and the other part is buried. Whereas in postmenopausal patients with malignant uterine tumor, trypotophan seems to be on the surface, exposed to absorbance. Hence progesterone receptors in postmenopausal patients with uterine cancer showed two trypotophan peaks.

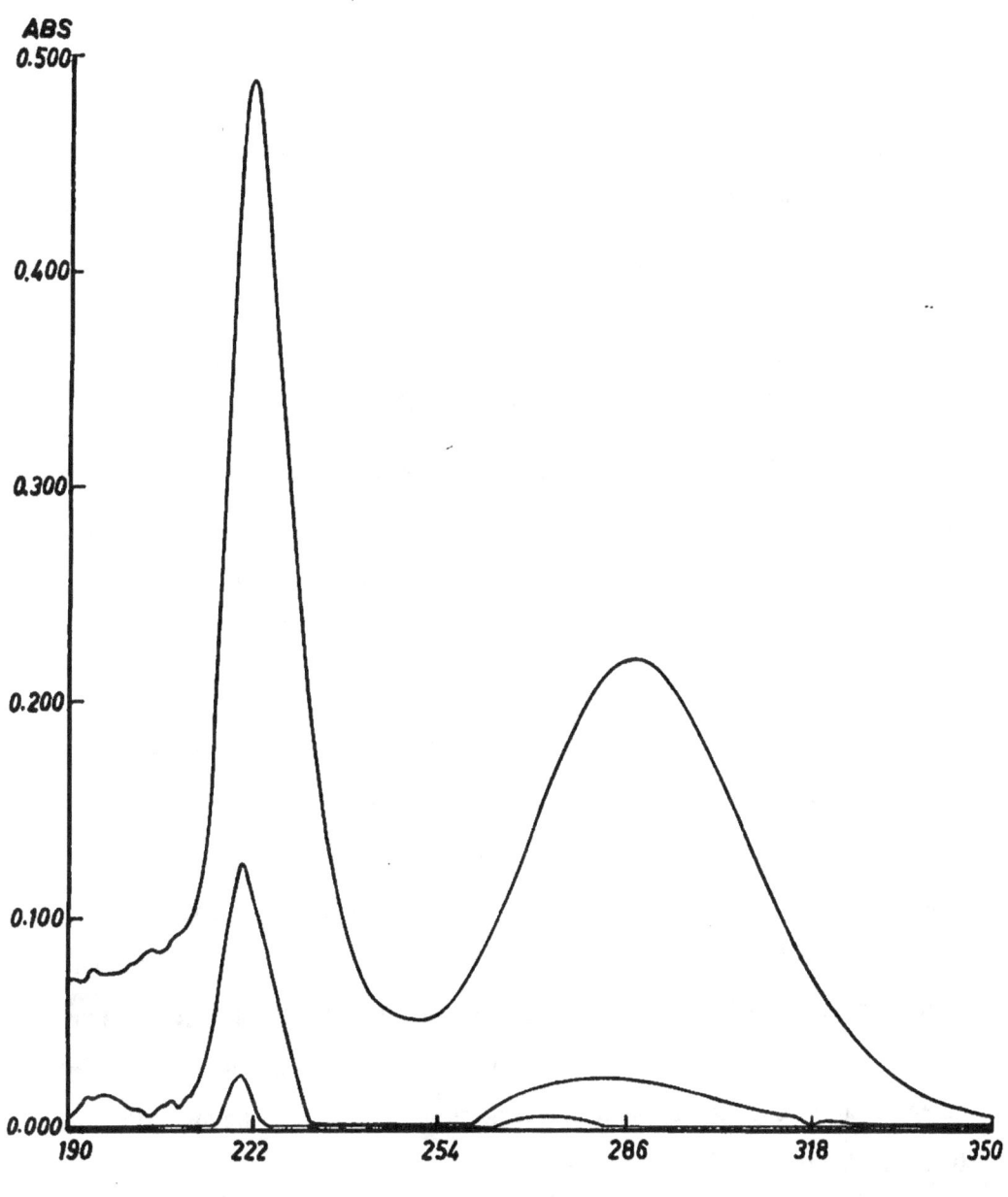

Figure(3-19) : The u.v spectrum of progesterone receptors. Details are described in section(2.2.7.1)

3.4.2 FACTORS AFFECTING THE ABSORPTION PROPERTIES OF PROGESTERONE RECEPTORS IN EACH OF PREMENOPAUSAL PATIENTS WITH BENIGN UTERINE TUMOR PRE AND POSTMENOPAUSAL PATIENTS WITH UTERINE CANCER

The absorption spectrum of any protein is primarily determined by the chemical structure of the protein. However, a large number of environmental factors products detect the changes in λ_{max}. Accordingly environmental factors such as the polarity of the solvent and the pH were studied.

3.4.2.1 PH EFFECT ON THE U.V SPECTRUM OF PROGESTERONE RECEPTORS

The pH of the solvent determines the ionization state of progesterone receptors. Table(3-8) shows the values of the λ_{max} for the progesterone receptors at different pH (7.0-8.6). In premenopausal patients with benign uterine tumor at pH from 7.0 to 7.4, one λ_{max} value was obtained; at 217.7nm which assigned to tryptophan. In premenopausal patients with uterine cancer at pH from 7.0 to 7.8, two λ_{max} values were obtained; at 288.3nm and 218.5nm these λ_{max} values are assigned to tryptophan. In postmenopausal patients with uterine cancer at pH from 7.0 to 7.6, one λ_{max} value was obtained; at 217.7nm which assigned to tryptophan. Increasing the pH to 8.6, new λ_{max} values were obtained in each of premenopausal patients with benign uterine tumor, pre and postmenopausal patients with malignant uterine tumor. This change could be attributed to the ionization of some amino acids that conform each protein molecule. It is suggested that each progesterone receptors have a different structure, due to the effect of pH.

Table(3-8): **The pH effect on the u.v spectrum of progesterone receptors.**

Details are described in section(2.2.7.2)

pH	Premenopausal patients with benign uterine tumor λ_{max} (nm)	Premenopausal patients with malignant uterine tumor λ_{max} (nm)	Premenopausal patients with malignant uterine tumor λ_{max} (nm)
7.0	217.7	288.3 , 218.5	217.7
7.4	216.3	288.3 , 218.5	216.3
7.6	220.6 , 203.8 , 194.2	286.9 , 219.2	226.3
7.8	286.9 , 221.3 , 203.8	272.6 , 218.5	280.5 , 291.9
8.2	274.8 , 217.7 , 206.3	286.9 , 219.2 , 193.2	278.3 , 217.7 , 197.1
8.6	283.3 , 216.3	286.9 , 219.2	284.7 , 217.0

3.4.2.2 POLARITY EFFECTS ON THE U.V SPECTRUM OF PROGESTERONE RECEPTORS

3.4.2.2.1 EFFECT OF 10% ETHANOL ON THE PROGESTERONE RECEPTORS SPECTRUM

Table(3-9) shows the λ_{max} values of progesterone receptors at two different pH (7.6, 8.6). In premenopausal patients with benign uterine tumor at pH 7.6, showed a different λ_{max} value; at 210.4nm. This λ_{max} value is assigned to histidine. In premenopausal patients with malignant uterine tumor at pH 7.6, λ_{max} values; showed a shift at a shorter wavelength. These wavelengths at 289.1 and 218.0nm are assigned to tryptophan. In postmenopausal patients with malignant uterine tumor at pH 7.6, a new λ_{max} value is obtained at 208.5nm, which assigned to phenylalanine.

The new λ_{max} value that appeared in premenopausal patients with benign uterine tumor and postmenopausal patients with malignant uterine tumor in presence of 10% ethanol, indicate that the protein was defolded due to change in the secondary and tertiary structure of the protein that bring the histidine and phenylalanine, respectively to expose to absorbance. In premenopausal patients with malignant uterine tumor, the shift in λ_{max} value, may be indicate that the protein is sensitive to change in the polarity of the solvent, the amino acid showing this change in the λ_{max} value may be on the surface of the protein.

Table (3-9): The effect of 10% ethanol on progesterone receptors. Details are described in section(2.2.7.2)

Samples	pH	10% ethanol λ_{max} (nm)
Premenopausal patients with benign uterine tumor	7.6 8.6	210.4 263.8 218.0 205.9 197.5
Premenopausal patients with malignant uterine Tumor	7.6 8.6	289.1 218.0 287.3 219.1
Postmenopausal patients with malignant uterine Tumor	7.6 8.6	208.5 218.0 200.9

3.4.2.2.2 EFFECT OF 10% POLYETHYLENE GLYCOL ON THE PROGESTERONE RECEPTORS

Table (3-10) shows the λ_{max} values of progesterone receptors at pH 7.0. In premenopausal patients with benign uterine tumor, additional λ_{max} values are obtained; at 191.4, 195.3 and 199.6nm. These λ_{max} values are assigned to tyrosine. In premenopausal patients with malignant uterine tumor, additional λ_{max} value is obtained; at 195.3nm which is assigned to tyrosine also. In postmenopausal patients with malignant uterine tumor, λ_{max} value showed a shift at a shorter wavelength. This wavelength is located; at 215.6nm and assigned to tryptophan. The additional λ_{max} values that obtained at pH 7.0 in presence of 10% PEG in premenopausal patients with benign and malignant uterine tumor, could be attributed to a change in the protein structure that bring the tyrosine-residue to the surface of the protein. In postmenopausal patients with malignant uterine tumor, the shift in λ_{max} value may indicate that tryptophan showing a change in λ_{max} and may be on the surface of the protein molecule. Thus the change in λ_{max} value may indicate that the protein is sensitive to changes in the polarity of the solvent, which indicate that a certain amino acid may be on the surface of the protein.

Table(3-10): The effect of 10 % PEG on progesterone receptors. Details are described in section(2.2.7.2)

Samples	pH	10 % PEG λ_{max} (nm)
Premenopausal patient with benign uterine Tumor	7.0	219.2 199.6 195.3 191.4
Premenopausal patients with malignant uterine Tumor	7.0	288.3 217.7 195.3
Postmenopausal patients with malignant uterine tumor	7.0	215.6

3.4.2.2.3 EFFECT OF UREA ON THE PROGESTERONE SPECTRUM

Table(3-11) shows the λ_{max} values of progesterone receptors at two different pH (7.0, 8.6). In premenopausal patients with benign uterine tumor at pH 7.0, additional λ_{max} value is obtained; at 200.6nm, which is assigned to phenylalanine and a shift in the λ_{max} value at a shorter wavelength. This wavelength; at 215.6nm is assigned to tryptophan. In premenopausal patients with uterine cancer at pH 7.0, additional λ_{max} value is obtained; at 198.9nm which assigned to tyrosine and a shift in the λ_{max} value at a shorter wavelengths. These wavelengths; at 286.9 and 217.0nm are assigned to tryptophan. In postmenopausal patients with uterine cancer at pH 7.0, additional λ_{max} value is obtained; at 202.8nm which is assigned to phenylalanine and a shift in the λ_{max} value at a shorter wavelength. This wavelength; at 216.3nm which assigned to tryptophan. The additional values of λ_{max} and the shift in λ_{max} value to a shorter wavelength at pH 7.0, in premenopausal patients with benign uterine tumor, pre and postmenopausal patients with uterine cancer in the presence of 8mM urea, indicate that the protein was sensitive to changes in the polarity of the solvent, that bring the phenylalanine, tyrosine and phenylalanine respectively to the surface of each protein molecule.

Table(3-11): The effect of 8mM Urea on progesterone receptors. Details are described in section(2.2.7.2)

Samples	pH	8mM urea λ_{max} (nm)
Premenopausal patients with benign uterine tumor	7.0	215.6
		200.6
	8.6	265.5
		222.0
Premenopausal patients with malignant uterine tumor	7.0	286.9
		217.0
		198.9
	8.6	287.6
		219.0
Postmenopausal patients with malignant uterine tumor	7.0	216.3
		202.8
	8.6	217.7
		206.3
		198.9

3.4.3 PH TITRATION

Figure(3-20) shows the tryptophan numbers that located on the surface of the receptor in premenopausal patients with benign uterine tumor and postmenopausal patients with malignant uterine tumor were approximately equal, but in premenopausal patients with malignant uterine tumor 40% of tryptophan numbers was on the surface and 60% inside the receptor molecule[132].

It is suggested that progesterone receptors in each of premenopausal patients with benign uterine tumor, per and postmenopausal patients with malignant uterine tumor have a different structure .

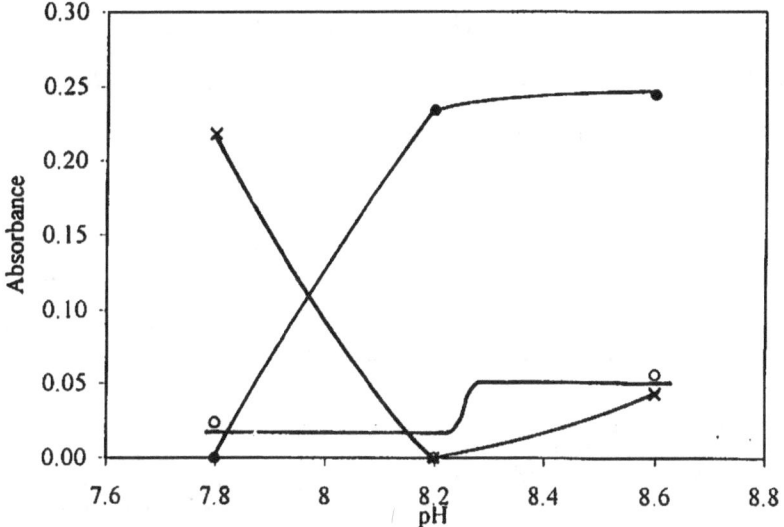

Figure(3-20): pH titration in three groups of uterine tumors
(•)Malignant tumors of premenopausal patients
(o)Malignant tumors of postmenopausal patients
(x) Benign tumors of premenopausal patients
Details are described in section (2.2.7.3).

3.4.4 SOLVENT-PERTURBATION ON PROGESTERONE RECEPTORS IN EACH OF PREMENOPAUSAL PATIENTS WITH BENIGN UTERINE TUMOR, PRE AND POSTMENOPAUSAL PATIENTS WITH MALIGNANT UTERINE TUMOR

Table(3-12) shows the λ_{max} values, absorbance, the change in the absorbance, in the presence and absence of different perturbants (25% PEG, 8mM urea and 8mM urea with 0.03M KCl) at pH 7.0. The effect of different perturbants, caused a decrease in λ_{max} value; at 217.7nm in premenopausal patients with benign uterine tumor, at 288.3nm and 218.5nm in premenopausal patients with malignant uterine tumor and; at 217.7nm in postmenopausal patients with malignant uterine tumor. These λ_{max} values are assigned to tryptophan-residue. These results indicate that the protein structure was affected in the presence of such perturbant. Accordingly, if the number of tryptophan-residues is known, it is possible to determine the tryptophan numbers of these residues either are internal or external in a protein molecule.

Table(3-12) : Solvent perturbation on progesterone receptors in premenopausal patients with benign uterine tumor, pre and postmenopausal patients with malignant uterine tumor. Details are described in section(2.2.7.4)

1- 25% (w/v) PEG at pH = 7.0

Samples	without PEG		With PEG		
	λ_{max} (nm)	A°	λ_{max} (nm)	A°	ΔA°
Premenopausal patients with benign uterine tumor	217.7	0.177	215.6	0.306	0.129
Premenopausal patients with uterine cancer	288.3	0.083	286.9	0.196	0.113
	218.5	0.315	217.7	0.706	0.391
Postmenopausal patients with uterine cancer	217.7	0.166	215.6	0.238	0.072

2- 8mM urea at pH = 7.0

Samples	without urea		With urea		ΔA°
	λ_{max} (nm)	A°	λ_{max} (nm)	A°	
Premenopausal patients with benign uterine tumor	217.7	0.177	215.6	0.201	0.024
Premenopausal patients with uterine cancer	288.3	0.083	286.9	0.107	0.024
	218.5	0.315	217.0	0.423	0.108
Postmenopausal patients with uterine cancer	217.7	0.166	216.3	0.184	0.018

3-8mM urea and 0.03M KCl at pH = 7.0

Samples	without urea & KCL		With urea & KCL		ΔA°
	λ_{max} (nm)	A°	λ_{max} (nm)	A°	
Premenopausal patients with benign uterine tumor	217.7	0.177	216.3	0.185	0.008
Premenopausal patients with uterine cancer	288.3	0.083	286.9	0.100	0.017
	218.5	0.315	217.0	0.401	0.086
Postmenopausal patients with uterine cancer	217.7	0.166	215.6	0.168	0.002

Conclusions

1- A higher incidence of progesterone receptors was obtained in malignant than in benign uterine tumors. Also a higher incidence of progesterone receptors was obtained in premenopausal patients with malignant uterine tumors than in postmenopausal patients with malignant uterine tumors.

2- The developed protocol for the assay of progesterone receptors is capable to analyze these receptors and the procedure is suitable for the assessment of progesterone receptors in benign and malignant uterine tumors.

3- The kinetic studies of the ^{125}I-progesterone binding to its receptors in premenopausal patients with uterine cancer showed that the binding reaction is a temperature and time dependent process. The result indicate that the reaction is second order at (4, 10, 25, 37°C).

4- The results obtained from the thermodynamic studies on the association of progesterone with its receptors, indicate that the binding reaction was entropically driven $(\Delta S° > 0)$.

5- The spectroscopic studies on progesterone receptors in premenopausal patients with benign uterine tumors pre- and postmenopausal patients with malignant uterine tumors revealed a characteristic spectrum for each receptor.

Future Work

1- Purification of the uterine receptors in different groups of tumors.

2- Molecular Characterization of those receptors of uterine tumors.

3- Molecular biology of uterine tumors receptors.

4- Correlation of these receptors of uterine tumors with those of other specific tumor markers.

References

1-Porth C. M. Structure and Function of the Female Reproductive System. *Pathopgysiology Concepts of Altered Health States*. 4th. ed. J. B. Lippincott Company. 1995, pp 747.

2-del Regato J. A., Spjut H. J. Female genital organs / Endometrium. *Cancer Diagnosis, Treatment, and Prognosis*. The C. V. Mosby Company. 5th. ed. 1977 pp 743.

3-Barnes J. The Uterus. *Lecture Notes on Gynaecology*. 5th. ed. Blakwell Scientific Publications. 1983, pp 6.

4-Tortora G. J., Anagnostakos N. P. The Reproductive Systems. *Principles of Anatomy and physiology*. 4th. ed. Harber International Edition. 1984, pp 713.

5-Porth C. M. Altrations in Structure and Function of the Female Reproductive System. *Pathopgysiology Concepts of Altered Health States*. 4th. ed. J. B. Lippincott Company. 1995, pp 761.

6-Barnes J. Infection-Vaginitis and Cervicitis. *Lecture Notes on Gynaecology*. 5th. ed. Blakwell Scientific Publications. 1983, pp 55.

7-Barnes J. Female Genital Tuberculosis. *Lecture Notes on Gynaecology*. 5th. ed. Blakwell Scientific Publications. 1983, pp 68.

8-Walter G. B., Israel M. S. Structure and Effects of Sum commn Tumors. *General Pathology*. 6th. ed. International Student Edition. 1988, pp 345.

9-Dotters D. J. and Droegemueller W. In: *Current Therapy*. edited by Rakel R.B. Baylor College of Medicine, Houston, Texas.1986, pp 881.

10-Barnes J. Benign Tumors of the Uterus. *Lecture Notes on Gynaecology*. 5th. ed. Blakwell Scientific Publications. 1983, pp 71.

11-Newton J. R. Tumors. *A Pocket Gynaecology*. 9th. ed. Churchill LivingStone. 1979, pp 66.

12-Chervenak F. A. Uterine Fibroids.*The Female Patient Guide*.1996, 21:43.

13-Schroeder S. A., Tierney L. M., et al. *Current Medical Diagnosis and Treatment* . 1989, pp 283.

14-Goldzieher JW, Maqueo M., Ricaud L, et al. *Am J Obstet Gynecol*. 1966, 96:1078.

15-Barnes J. Essential Gynaecological Endocrinology. *Lecture Notes on Gynaecology*. 5th. ed. Blakwell Scientific Publications. 1983, pp 190.

16-Maheux R. *Horm Res*. 1989; 32 (Suppl 1):125.

17-Barnes J. Malignant Tumors of the Uterus. *Lecture Notes on Gynaecology*. 5th. ed. Blakwell Scientific Publications. 1983, pp 82.

18-Homesley HD, Zaino R.. *Semin Oncol*. 1994, 21:71.

19-Parazini F., La Vecchia C., Bocciolone L., Franceschi S. *Gynecol Oncol.* 1991, 41:1.

20-Pernoll M. L., Benson R. C. *Current Obstetric and Gynecologic Diagnosis and Treatment.* 6th. ed. Middle East Edition. 1987, pp 882, 905.

21-Ehrlich C. E., Young P. C. M., Cleary R. E. *Am G Obstet Gyneo.* 1981, 141:539.

22-Burke TW, Munkarah A., Kavanagh JJ, et al. *Gynecol Oncol.* 1993, 51:397.

23-Burke T. W., Tortoleroluna G., Malpica A., Vicki V., et al. *Gynecologic Cancer Prevention.* 1996, 23 (2):411.

24-Schwartz M. K. Labrotory Aid to Diognosis. Enzymes. *Cancer.* 1993, 37.

25-Pandha H. S., Waxman J., Sikora K. *British Journal of Hospital Medcine.* 1994, 51 (6):297.

26-Bates SE and Longo DL. *Semin Oncol.* 1987, XIV (2):102.

27-Van Nagell J. R., Donaldson E. S., Hanson M. B., Gay E. C. and Pavlik E.J. *Cancer.* 1981, 48:495.

28-Porth C. M. Alterations in Cell Differenitiation : Neoplasia. *Pathopgysiology Concepts of Altered Health States.* 4th. ed. J. B. Lippincott Company. 1995, pp 87.

29-Statland BE. *Diagn Med.* 1981, 4:2.

30-Niman HL. *J Clin Lab Anal.* 1987, 1:28.

31-Pastore M. and Francioni S. *J Nucl Med Allied Sci.* 1989,33(3 suppl):107.

32-Thomas CMG, Segers MFG and Houx PEW. *Ann Clin Biochem.* 1985, 22:236.

33-Murray R. K., Mayes P. A., Granner D. K., and Rodwell V. W. *Harper's Biochemistry.* 22th. ed. Prentice-Hall. International Inc. 1993, pp 650.

34-Mackay E. V., Khoo S. K. and Daunter B. *Tumor Markers in Gynecology Oncology.* 2nd. ed. (M Coppleson, E. d.). 1987, pp 270.

35-Gold P. and Freedman S. O. *J Exp Med.* 1965, 122:439.

36-Khoo S. K., Warner N. L., Lie J. T., and Mackay I. R. *Int J Cancer.* 1973, 11:681.

37-Crum G. P. and Fenoglio C. M. *Diagn Gynecol Obstet.* 1980, 2 (3) :103.

38-Ludwig H. *Eur J Obstet Gynaecol Reprod Biol.* 1988, 28 :104.

39-Sikorska H., Shuster J. and Gold P. *Clin Detect Prevent.* 1988, 12 :321.

40-Jullienne A., Calmettes C., Moukhtar M. S. and Milhaud G. *Oncodev Biol Med.* 1980, 1:137.

41-Galen R. S. *Am Clin Lab Sci.* 1977, 7:51.

42-Pustaszeri G. and Mach J. P. *Immunochem.* 1973, 10:197.

43-Van Nagell JR, Meeker W., Parker JC, Harralson JD. *Cancer.* 1975, 35:1372.

44-Donaldosn E. S., Van Nagell J. R., et al. *Cancer.* 1980, 45:948.

45-Nouwen E. J., Pollet D. E., Ecrdekyns M. W., Hedrix P. G., Briers T. W. and De Brone M. E. *Cancer Res*. 1986, 46:866.

46-Masahashi T., Matsuzawa K. Ohsawa M., Narita O., Asai T. and Ishihara M. *Obstet gynecol*. 1988, 72 (3):328.

47-Poels L. G., Peters D. and Van Megen Y. *J Natl Cancer Inst*. 1986, 76 (5): 781.

48-Altaras M. M., Goldberg G. I., Levin W., Bloch B., Darge I. and Smith J. A. *Gynecol Oncol*. 1988, 30:26.

49-Alvarez R. D., Boots L. R., Shingleton H. M., Hatch K. D., Hubbard J., Soong S. J. and Potter M. E. *Gynecol Oncol*. 1987, 26:284.

50-Vanderburg M. E. L. *Gynocol Oncol*. 1988, 30:307.

51-Staratton J. A. S., Rettenmaier M. A., Phillips H. B., Herbabutya S. and Di Saia P. J. *Obstet gynecol*. 1988, 71 (1):20.

52-Zurawski V. R., Knapp R. C., Einhorn N., Kenemans P., Mortel R. and Ohmi K. *Gynecol Oncol*. 1988, 30:7.

53-Daunter B. *Gynecol Oncol*. 1990, 39:1.

54-Maughaw T. S., Fish R. G., Shelley M., Jasani B., Williams G. T. and Adams M. *Gynecol Oncol*. 1988, 30:342.

55-Scambia G. Benedetti P., et al. *Gynecol Oncol*. 1988, 30:265.

56-Scambia G. Benedetti P., et al. *Gynecol Oncol*. 1988, 45:263.

57-Panici P. B., Scambia G., et al. *Gynecol Obstet Invest*. 1989, 27:208.

58-Ward BG, McGuckin MA, et al. *Cancer*. 1993, 71:430.

59-Kiein F. A. *Urol Clin North Am*. 1993, 20:67.

60-Sell S. *Clin Lab Med*. 1990, 10:1.

61-Donaldson ES, Van Nagell JR., Gay EC, et al.*Cancer*. 1980, 45:948.

62-Malkasian G. D., Podratz C., Stauhope C. R. and Podratz K. C. *Am J Obstet Gynecol*. 1986, 155 (3):515.

63-Kabawat S. E., Bast R. C., Bhan A. K., Welch W. R., Knapp R. C. and Colvin R. B. *Int J Gynecol Pathol*. 1983, 2:275.

64-Duk JM, Aalders JG, Fleuren GT, et al. *Am J Obstet Gynecol*. 1986, 155:1097.

65-Patsner B., Mann WJ, Cohen H., et al. *Gynecol Oncol*. 1988, 29:131.

66-Patsner B., Orr JW, Mann WJ, et al. *Am J Obstet Gynecol*. 1990, 162:427.

67-Schwartz PE, Chambers SK, Chambers JT, et al. *Cancer*. 1987, 60:353.

68-Burke T. W., Tortolero-Luna G., Malpica A., et al. *Gynecologic Cancer Prevention*. 1996, 23 (2):411.

69-Stefanini M. *Cancer*. 1985, 55:1931.

70-Schapira F. *Adv Cancer Res*. 1973, 18:77.

71-Hayashi-S. *Acta Obstet Gynaecol Jpn*. 1981, 33 (7):1035.

72-Kalpaxis D. L. and Giannoulaki E. E. *Clin Chem*. 1989, 35 (5):844.

73-Vergnon JM, Guidollet J, Gateau O., et al. *Cancer*. 1984, 54:507.

74-Sudo K., Maekawa M., Watanabe H., et al. *Clin Chem*. 1986, 32:1420.

75-Giannoulaki EE, Kalpaxis DL, Tentas C., Fessas F. *Clin Chem.* 1989, 35:396.

76-Geyer H., Afting EC., Toussi P. *Arch Cynecol.* 1984, 234 (3):181.

77-Reddy VV, Rose LL. *Am J Obstet Gynecol.* 1979, 135:415.

78-Pollow K., Sinnecker E. and Borquoi B. *J Clin Chem Clin Biochem.* 1978, 16:493.

79-Lee YS, Raju GC. *J Pathol.* 1988, 155 (3):201.

80-Holinka CF, Gurpide E. *In Vitro Cell Dev Biol.* 1985, 21 (12):697.

81-Sugawara S. *Nippon-Sanka-Fujinka-Gakkai-Zasshi.* 1981, 33 (10):1749.

82-Vaitukaitis JL, Braunstein GP, et al. *Am J Obstet Gynecol.* 1972, 113:751.

83-Kaplan L. A., Pesce A. J. The Gonads. *Clinical Chemictry Theory, Analysis and Correlation.* 2nd. ed. The C. V. Mosby Company. 1989, pp 655.

84-Lazar M. A. *Endocrinology and Metabolism Clinics of North American.* 1991, 30 (4):681.

85-Kumar P. and Clark M. Endrocrinology clinical Medicine. *A Textbook for Medical Students and Doctors.* 3rd. ed. Butler and Tanner Ltd, Frome and London. 1994, pp 770.

86-Speroff L., Glass R., Kase N. *Clinical Gynecologic Endocrinology and Infertilit.* 4th. ed. 1989, pp 6.

87-Ganong W. F. The Gonads : Development and Function of the Reproductive System. *Review of Medical Physiology.* 5th. ed. Long Medical Book. 1993, pp 417.

88-Nelson L. M. *The Female Patient.* 1992, 17:15.

89-Lupulescu A. P. *Cancer.* 1996, 78 (11):2264.

90-Jensen EV, Desombre ER. *Science.* 1973, 182:126.

91-Kerr JF, Winterford CM, Harmon BV. *Cancer.* 1994, 73:2013.

92-Creasman W. T., Soper J. T., et al. *Am J Obstet Gynecol.* 1985, 151:922.

93-Creasman WT, McCarty KS, Barton TF, et al. *Obstet Gynecol.* 1980, 55:363.

94-Kauppila A., Kujansuu E., Vihko R. *Cancer.* 1982, 50:2157.

95-Martin JD, Hahnel R., McCartney AJ, Woodings TL. *Am J Obstet Gynecol.* 1983, 147:322.

96-Ehrlich C. E., Young P. C. M., Stehman F. B., Sutton G. P. and Alford W. M. *Am J Obstet Gynecol.* 1988, 158:796.

97-Palmer D. C., Muir I. M., Alexander A. I. et al. *Obstet Gynecol.* 1988, 72:388.

98-McGuire J. L. and DeDella C. *Endocrinology.* 1971, 88:1099.

99-Verma U. and Laumas K. R. *Biochim Biophys Acta.* 1973, 317:403.

100-Toft D. O. and O'Malley B. W. *Endocrinology.* 1972, 90:1041.

101-Faber L. E., Sandmann M. L. and Stavely H. E. *J Biol Chem.* 1972, 247:5648.

102-Corvol P., Falk R., Freifeld M. and Bardin C. W. *Endocrinology*. 1972, 90:1464.

103-Kontula K., Janne O., Luukkainen T. and Vihko R. *Biochim Biophys Acta*. 1973, 328:145.

104-Pollow K. In : *Hormones in Normal and Abnormal Human Tissues*. de Gruyter W. Berlin. New York. 1981, pp 378.

105-Wiest W. G., Rao B. R. *Adv BioSci*. 1970, 7:251.

106-O'Malley B. W. and Means A. R. *Science*. 1974, 183:610.

107-Pollow K. Schmidt-Gollwitzer M., Nevinny-Stickel J. : Progesterone Receptors in normal human endometrium an Endometrial Carcinoma. In : *"Progesterone Receptors in Normal and Neoplastic Tissues"*. Eds. McGuire W. L. Raynaud J.-P., Baulieu E.-E. Raven Press. New York. 1977, pp 313.

108-Bayard F., Damilano S., Robel P., Baulieu E.-E. *J Clin Endocrinol Metab*. 1978, 46:635.

109-Soper JT, Christensen CW. *Clin Obstet Gynecol*. 1986, 13:825.

110-Quinn MA, Cauchi M. Fortune D. *Gynecol Oncol*. 1985, 21:314.

111-Benraad TJ, Friberg LG, Koenders AJM, Kullander S. *Acta Obstet Gynecol Scand*. 1980, 59:155.

112-Gowenlock AH. editor. *Varley's Practical Clinical Biochemistry*. 6th. ed. Heinemann Medical Books. London. 1988, pp 1018.

113-Bounaud JY, Bounaud MP, Metayer T., et al. *Eur J Cancer*. 1988, 24:461.

114-Lowry OH, Rosebrough NJ, Farr AL, et al. *J Biol Chem*. 1951, 93:265.

115-Morris BJ. *Clinica Chimica Acta*. 1976, 73:213.

116-Scatchard G. *Ann NY Acad Sci*. 1949, 51:660.

117-Vermeulen A., Verdonck L. *Clin Endocrinol*. 1978, 9:59.

118-Soules M. R., McCarty K. S. *Am J Obstet Gynecol*. 1982, 143:6.

119-Bayard F., Damilano S., Robel P. and Baulieu E-E. *Journal of Clinical Endocrinology and Metabolism*. 1978, 46 (4):635.

120-Lantta M., Karkkainer. J. and Lehtovirta P. *Am J Obstet Gynecol*. 1983, 147 (15):627.

121-Daxembichler G., Grill H. J. Wiesinger H., Wittliff J. L. and Dapunt O. In: *Multiple Molecular Forms of Steroid Hormone Receptors*. Agarwal MK. editor. Elsevier, North-Holland Biomedical Press.1977, pp. 163.

122-Buller R. E., Toft D. O., Schrader W. T. and O'Malley B. W. *The Journal of Biological Chemistry*. 1975, 250 (3):801.

123-Smith H. E., Smith R. G., Toft D. O., Neergaard J. R., Burrows E. P. and O'Malley B. W. *the Journal of Biological Chemistry*. 1974, 249 (18):5924.

124-Seeley DH, Wang WY, Salhanick HA. *Biochim Biophys Acta*. 1980, 632 (4):536.

125-Weiland GA and Molinoff C. *Life Sci.* 1981, 29:313.

126-Segel IH. *Biochemical Calculations.* 2nd ed. John Willey and Sons, Inc. 1976, pp 311.

127-Nemethy G and Scheraga HA. *J phys chem.* 1962, 66:1773.

128-Waelbroeck M., Van Obberghen E. and DeMeyts P. *J Biol chem.* 1979, 254:7736.

129-Ross PD and Subramanian. *Biochemistry.* 1981, 20:3096.

130-Blumenthal D. K. and Stull J. T. *Biochemistry.* 1982, 21:2386.

131-Laporte D. C., Wierman B. M. and Storm D. R. *Biochemistry.* 1980,19:3814

132-Freifelder D. *physical biochemistry.* 2nd. ed.1982, pp 512.

www.ingramcontent.com/pod-product-compliance
Lightning Source LLC
Chambersburg PA
CBHW080825180526
45168CB00006B/2577